W9-AGB-722

α Alpha

Single-Digit Addition and Subtraction

Instruction Manual

By Steven P. Demme

1-888-854-MATH (6284) - mathusee.com
sales@mathusee.com

Alpha Instruction Manual: Single-Digit Addition and Subtraction

Published and distributed by Demme Learning

mathusee.com

1-888-854-6284 or +1 717-283-1448 | demmelearning.com
Lancaster, Pennsylvania USA

ISBN 978-1-60826-079-9
Revision Code 1118-B

Printed in the United States of America by Innovative Technologies in Print
2 3 4 5 6 7 8 9 10

For information regarding CPSIA on this printed material call: 1-888-854-6284
and provide reference #1118-060820

Building Understanding in Teachers and Students to Nurture a Lifelong Love of Learning

At Math-U-See, our goal is to build understanding for all students.

We believe that education should be relevant, skills-based, and built on previous learning. Because students have a variety of learning styles, we believe education should be multi-sensory. While some memorization is necessary to learn math facts and formulas, students also must be able to apply this knowledge in real-life situations.

Math-U-See is proud to partner with teachers and parents as we use these principles of education to **build lifelong learners.**

Curriculum Sequence

$\int$	**Calculus**
cos	**PreCalculus** with Trigonometry
xy	**Algebra 2**
Δ	**Geometry**
x^2	**Algebra 1**
x	**Pre-Algebra**
ζ	**Zeta** Decimals and Percents
ε	**Epsilon** Fractions
δ	**Delta** Division
γ	**Gamma** Multiplication
β	**Beta** Multiple-Digit Addition and Subtraction
α	**Alpha** Single-Digit Addition and Subtraction
P	**Primer** Introducing Math

Math-U-See is a complete, K-12 math curriculum that uses manipulatives to illustrate and teach math concepts. We strive toward "Building Understanding" by using a mastery-based approach suitable for all levels and learning preferences. While each book concentrates on a specific theme, other math topics are introduced where appropriate. Subsequent books continuously review and integrate topics and concepts presented in previous levels.

Where to Start
Because Math-U-See is mastery-based, students may start at any level. We use the Greek alphabet to show the sequence of concepts taught rather than the grade level. Go to mathusee.com for more placement help.

Each level builds on previously learned skills to prepare a solid foundation so the student is then ready to apply these concepts to algebra and other upper-level courses.

Major concepts and skills for Alpha:

- Understanding place value
- Extending the counting sequence
- Fluently adding all single-digit numbers
- Solving for an unknown addend
- Understanding the relationship between addition and subtraction
- Fluently subtracting all single-digit numbers

Additional concepts and skills for Alpha:

- Telling and writing time by hours and minutes
- Recognizing and drawing simple geometric shapes
- Measuring length by repeating units
- Introducing halves and fourths
- Counting by 2s, 5s, 10s, and 100s
- Reading, writing, and interpreting word problems

Contents

HOW TO USE

Five Minutes for Success

Welcome to *Alpha*. I believe you will have a positive experience with the unique Math-U-See approach to teaching math. These first few pages explain the essence of this methodology, which has worked for thousands of students and teachers. I hope you will take five minutes and read through these steps carefully.

The student should be able to count and write the numbers from zero to nine and be ready for formal schooling.

If you are using the program properly and still need additional help, you may visit us online at mathusee.com or call us at 888-854-6284. **–Steve Demme**

The Goal of Math-U-See

The underlying assumption or premise of Math-U-See is that the reason we study math is to apply math in everyday situations. Our goal is to help produce confident problem solvers who enjoy the study of math. These are students who learn their math facts, rules, and formulas and are able to use this knowledge to solve word problems and real-life applications. Therefore, the study of math is much more than simply committing to memory a list of facts. It includes memorization, but it also encompasses learning the underlying concepts of math that are critical to successful problem solving.

Support and Resources

Math-U-See has a number of resources to help you in the educational process.

Many of our customer service representatives have been with us for over 10 years. They are able to answer your questions, help you place your student in the appropriate level, and provide knowledgeable support throughout the school year.

Visit mathusee.com to use our many online resources, find out when we will be in your neighborhood, and connect with us on social media.

More than Memorization

Many people confuse memorization with understanding. Once while I was teaching seven junior high students, I asked how many pieces they would each receive if there were fourteen pieces. The students' response was, "What do we do: add, subtract, multiply, or divide?" Knowing how to divide is important; understanding when to divide is equally important.

The Suggested 4-Step Math-U-See Approach

In order to train students to be confident problem solvers, here are the four steps that I suggest you use to get the most from the Math-U-See curriculum.

Step 1. Prepare for the lesson
Step 2. Present and explore the new concept together
Step 3. Practice for mastery
Step 4. Progress after mastery

Step 1. Prepare for the lesson

Watch the video lesson to learn the new concept and see how to demonstrate this concept with the manipulatives when applicable. Study the written explanations and examples in the instruction manual.

Step 2. Present and explore the new concept together

Present the new concept to your student. Have the student watch the video lesson with you, if you think it would be helpful. The following should happen interactively.

a. **Build:** Use the manipulatives to demonstrate and model problems from the instruction manual. If you need more examples, use the appropriate lesson practice pages.

b. **Write:** Write down the step-by-step solutions as you work through the problems together, using manipulatives.

c. **Say:** Talk through the why of the math concept as you build and write.

Give as many opportunities for the student to "Build, Write, Say" as necessary until the student fully understands the new concept and can demonstrate it to you confidently. One of the joys of teaching is hearing a student say, *"Now I get it!"* or *"Now I see it!"*

Step 3. Practice for mastery

Using the lesson practice problems from the student workbook, have students practice the new concept until they understand it. It is one thing for students to watch someone else do a problem; it is quite another to do the same problem themselves. Together complete as many of the lesson practice pages as necessary (not all pages may be needed) until the student understands the new concept, demonstrating confident mastery of the skill. Remember, to demonstrate mastery, your student should be able to teach the concept back to you using the Build, Write, Say method. Give special attention to the word problems, which are designed to apply the concept being taught in the lesson. If your student needs more assistance, go to mathusee.com to find review tools and other resources.

Step 4. Progress after mastery

Once mastery of the new concept is demonstrated, advance to the systematic review pages for that lesson. These worksheets review the new material as well as provide practice of the math concepts previously studied. If the student struggles, reteach these concepts to maintain mastery. If students quickly demonstrate mastery, they may not need to complete all of the systematic review pages.

You may use the application and enrichment pages (2012 student workbook) for additional practice and for variety.

Now you are ready for the lesson tests. These were designed to be an assessment tool to help determine mastery, but they may also be used as extra worksheets. Your student will be ready for the next lesson only after demonstrating mastery of the new concept and maintaining mastery of concepts found in the systematic review worksheets.

Tell me, I forget. Show me, I understand. Let me do it, I remember.
–Ancient Proverb

To this Math-U-See adds, *"Let me teach it, and I will have achieved mastery!"*

Length of a Lesson

How long should a lesson take? This will vary from student to student and from topic to topic. You may spend a day on a new topic, or you may spend several days. There are so many factors that influence this process that it is impossible to predict the length of time from one lesson to another. I have spent three days on a lesson, and I have also invested three weeks in a lesson. This experience occurred in the same book with the same student. If you move from lesson to lesson too quickly without the student demonstrating mastery, the student will become overwhelmed and discouraged as he or she is exposed to more new material without having learned previous topics. If you move too slowly, the student may become bored and lose interest in math. I believe that as you regularly spend time working along with the student, you will sense the right time to take the lesson test and progress through the book.

By following the four steps outlined above, you will have a much greater opportunity to succeed. Math must be taught sequentially, as it builds line upon line and precept upon precept on previously learned material. I hope you will try this methodology and move at the student's pace. As you do, I think you will be helping to create a confident problem solver who enjoys the study of math.

LESSON 1

Place Value and the Manipulatives

Two skills are necessary to function in the ***decimal system***: the ability to count from zero to nine and an understanding of ***place value.*** In the decimal system, where everything is based on ten (deci), we count to nine and then start over. To illustrate this, count the following numbers slowly: 800, 900, 1,000. We read these as eight hundred, nine hundred, one thousand. Now read these: 80, 90, 100. Notice how we count from one to nine and then begin again.

Alpha assumes that a child can count to nine. If you need to practice counting objects, be sure to arrange them in different patterns. For example, arrange the unit blocks in a circle as well as in a line. Once the child can count to nine in any arrangement, you can begin to work on place value.

Counting

On the number chart, we will begin with zero and proceed to nine. Traditionally we've started with one and counted to ten. Look at the two charts that follow and see which is more logical.

1	2	3	4	5	6	7	8	9	10
11	12	13	14	15	16	17	18	19	20
21	22	23	24	25	26	27	28	29	30

0	1	2	3	4	5	6	7	8	9
10	11	12	13	14	15	16	17	18	19
20	21	22	23	24	25	26	27	28	29

The second chart has only single-digit numbers in the first line. In the second line, each digit is preceded by a 1 in the tens place. The next line has a 2 preceding each digit instead of a 1. The first chart, though more familiar, has the 10, the 20, and the 30 in the wrong lines. When practicing counting, begin with zero, count to nine, and then start over.

Place Value

I define this important subject as "Every value has its own place." To an older child I would add, "Place determines value." Both are true. There are ten symbols to tell you how many and many values to represent what kind or what value. The numbers zero through nine tell us how many; ***units***, ***tens***, and ***hundreds*** tell us what kind. For the sake of accuracy, we will use the term "units" rather than "ones" to denote the first value. One is a counting number that tells us how many, and units is a place value that denotes what kind. This will save potential confusion when saying "ten ones" or "one ten." Remember, one is a number, and units is a place value. The numerals (0–9) tell us how many tens, how many hundreds, or how many units. We begin our study focusing on the units, tens, and hundreds, but there are other place values, such as thousands, millions, billions, and so on.

When teaching place value, I like to illustrate it by using a street, since I'm talking about a place. I call the street Decimal Street® and have the little green units house, the tall blue tens house next door, and the huge red hundreds castle beside the tens. We don't want to forget what we learned from counting—that we count only to nine and then start over. To make this more real, begin by asking, "What is the greatest number of units that can live in this house?" You can get any response to this question, from zero to nine, and you might say "yes" to all of them, but remind the student that the greatest number is nine. We can imagine how many little green beds, or green toothbrushes, or green chairs there would be in the house. Ask the student what else there would be nine of. Do the same with the tens and the hundreds. Remember that in these houses all the furniture will be blue (tens) or red (hundreds).

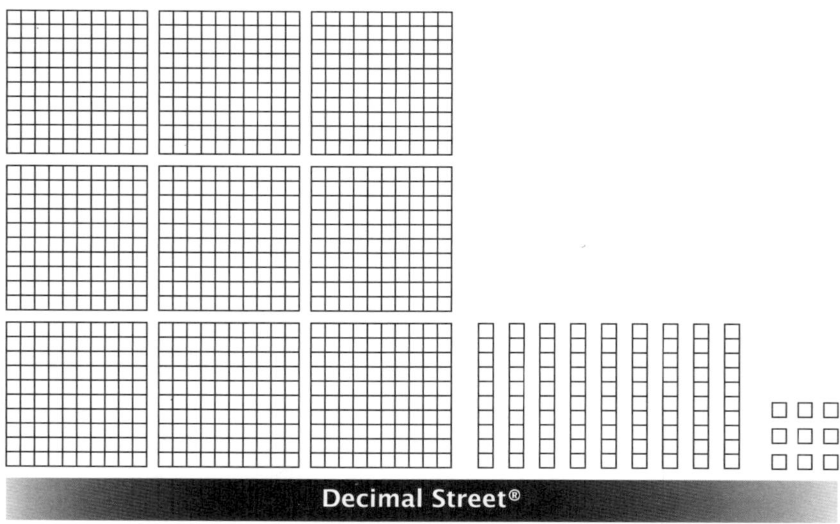

Throughout the program, whenever we teach, we will employ the following strategy: Build, Write, Say. To teach place value, we will first build the number and then count how many in each place. Finally, we write the number and read what we've written. There are directions for making your own Decimal Street® chart at the end of this lesson.

Let's build 142 (1 hundred, 4 tens, 2 units). Now count how many are "at home" at each house. I like to imagine going up to the door of each house and knocking to see how many are at home in each place. Write the digits 1 4 2 as you count (always beginning with the units) to show the value on paper. Then say, "One hundred, four tens, and two units, or one hundred forty-two." Build another example and have the students write how many are at home. When they understand this, write the number on paper and have them build it. Try 217. After they build it, read what they have built. Keep practicing, with you building and the student writing, and vice versa.

Here is another exercise to reinforce the fact that every value has its own place. I like to have the student close his/her eyes as I move the pieces around by placing the red hundreds where the units should be and vice versa. I then ask the student to make sure the blocks are all in the right place. You might call this "scramble the values" or "walk the blocks home." As the student looks at the problem and begins to work on it, I ask, "Is every value in its own place?"

Mr. Zero (0) is a very important symbol. He is a place holder. Let's say you were walking down Decimal Street and knocking on each door to see who was at home. If you knock on the Units door and three Units answer, you have three in the units place. Next door, at the Tens house, you knock, and no one answers. Yet you know someone is there feeding the goldfish and taking care of the bird, as a house sitter might when a family goes on vacation. Mr. Zero won't answer the door because he's not a Ten; he's the one who holds their place until the Tens come home. Upon knocking at the big red Hundreds castle, you find that two Hundreds answer the door. Thus, your numeral is 203.

Mention that even though we begin at the units end of the street and proceed right to left from the units to the hundreds, when we read numbers we do it from left to right. We want to get into the habit of counting units first so that when we add, we will add units first, then tens, and then hundreds.

Remember, teach with the blocks and then move to the worksheets once the student understands the new material.

You've probably noticed the important relationship between language and place value. Consider 142, read as one hundred forty-two. We know that it is made up of one red hundred square (one hundred), four blue ten bars (forty—*ty* for ten), and two units. The hundreds are very clear and self-explanatory, but the tens are where we need to focus our attention.

When pronouncing 90, 80, 70, 60, and 40, work on enunciating clearly so that 90 is ninety, not "ninedee." 80 is eighty, not "adee." When you pronounce the numbers accurately, not only will your spelling improve, but your understanding of place value will improve as well. Seventy (70) is seven tens, and sixty (60) is six tens. Forty (40) is pronounced correctly but spelled without the *u.* Carrying through on this logic, 50 should be pronounced "five-ty" instead of fifty. Thirty and twenty are similar to fifty-- not completely consistent but close enough so we know what they mean.

The teens can be problematic. Some researchers believe that students in Japan and China have a better understanding of place value than students in Europe and the United States. One of the reasons for this is that in the Chinese and Japanese languages, the words for numbers are very regular, and the words for numbers greater than nine are built quite logically from the words for zero to nine. In contrast, there are a number of irregular words for numbers in English and in other European languages, and the English language in particular is very irregular in the words for eleven through nineteen.

To compensate for this, I'm suggesting a new way to read the numbers 10 through 19. You can decide whether this method reinforces the place value concept and restores logic and order to the decimal system. Ten is "onety," 11 is "onety-one," 12 is "onety-two," 13 is "onety-three," and so on. It is not that students can't say ten, eleven, and twelve, but learning this method enhances their understanding and makes math logical. Also, children often think it is fun.

When presenting place value or any other topic in this curriculum, model how you think as you solve the problems. When you as the teacher work through a problem with the manipulatives, do it verbally so that, as the student observes, he or she also hears your thinking process. Then record your answer.

Example 1

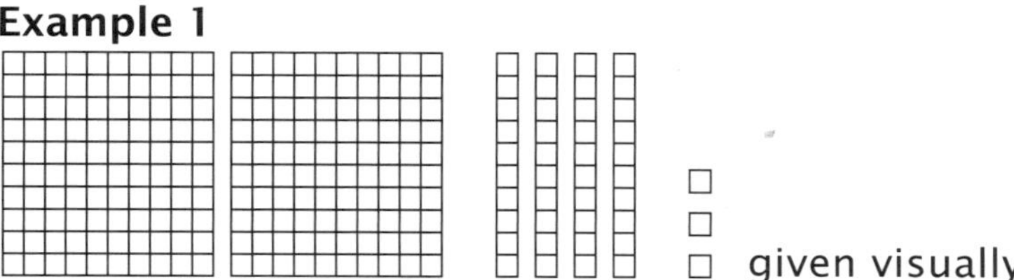

As you look at the picture, say the number slowly, going from left to right: "Two hundred forty-three." Then count, beginning with the units ("1-2-3") and write a 3 in the units place. Now count the tens ("1-2-3-4") and write a 4 in the tens place. Finally, count the hundreds ("1-2") and write a 2 in the hundreds place. Do several of these and then have the student try.

Many students have been taught to write an "open" four (4), while most printed material has a "closed" four (4). Either way is correct. Just be sure that the student recognizes both styles.

Example 2
274 (given in writing)

Read the number "two hundred seventy-four." Say "two hundreds" as you pick up two red hundred squares. Say "seven-ty, or seven tens" and pick up seven blue ten bars. Finally, say "four" and pick up four green unit pieces. Place them in the correct places as you repeat, "Every value has its own place."

Example 3
"one hundred sixty-five" (given verbally)

Read the number slowly. Say "one hundred" as you pick up one red hundred square. Say "six-ty, or six tens" and pick up six blue ten bars. Now say "five" and pick up five green unit pieces. Place them in the correct place as you say, "Every value has its own place." Finally, write the number 165.

Do several problems each way (in writing and verbally) and then give the student the opportunity to practice.

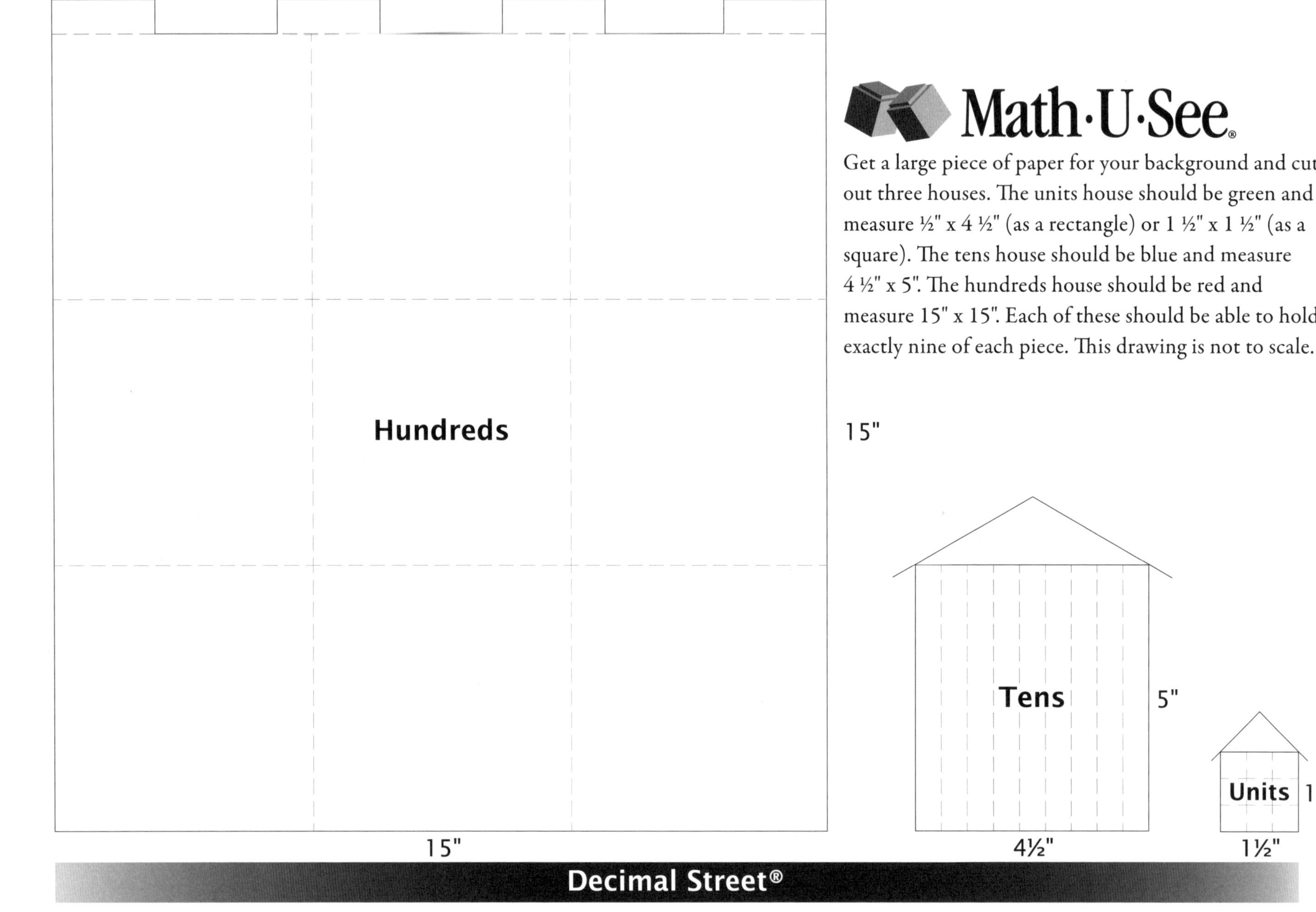
Math·U·See®
Get a large piece of paper for your background and cut out three houses. The units house should be green and measure ½" x 4 ½" (as a rectangle) or 1 ½" x 1 ½" (as a square). The tens house should be blue and measure 4 ½" x 5". The hundreds house should be red and measure 15" x 15". Each of these should be able to hold exactly nine of each piece. This drawing is not to scale.
Hundreds
15"
15"
Tens
5"
4½"
Units
1½"
1½"
Decimal Street®

Game for Place Value

Pick a Card - Make a set of cards with the numerals 0 through 9 written in green, one numeral on each card. Then make another stack of cards with the same numerals written in blue. (You may use the cards from the application and enrichment pages in the student workbook, if you wish.) Create one more stack of cards with the numerals 0 through 9 written in red. Shuffle the green cards; then have the child pick one and display that number of green unit blocks. For example, if the child picks a green 4, four green unit blocks should be counted out.

When the child is proficient at this game, try it with the blue cards and do the same thing, except choose the blue ten blocks instead of the green unit blocks. When he can do the tens well, use both sets of cards. Have the child choose one card from the green pile and one card from the blue pile and pick up the correct number of blue ten blocks and green unit blocks. When he or she is an expert at this, add the red cards and proceed as before. Shuffle each set and place them in three stacks. Have the student draw from each stack and show you with the blocks what number has been drawn.

Counting to 20

The transition from 9 to 10 is pivotal in understanding counting to 100 and later on for regrouping in addition and subtraction.

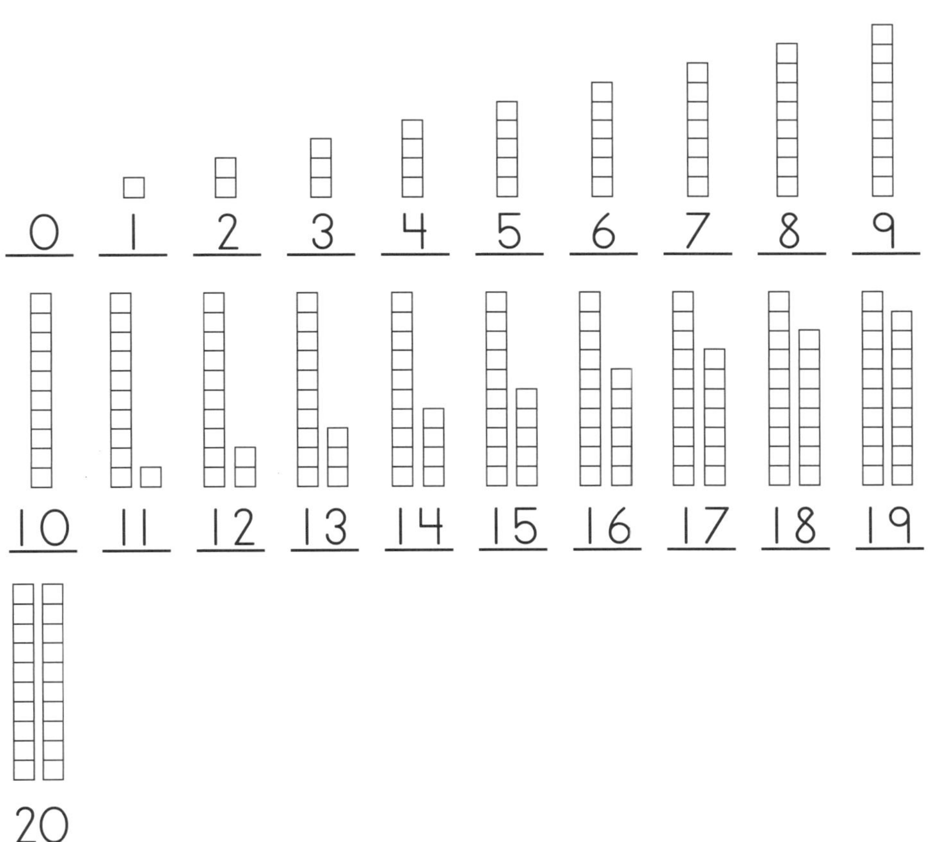

On the worksheets, the students will practice writing the numbers on the lines. I left in a few numbers to make sure your student doesn't miss any.

___ 1 ___ 3 4 ___ ___ 7 ___ 9

10 ___ ___ ___ ___ 15 16 ___ 18 ___

20

The writing exercises show open fours, while closed fours are used in most problems on the worksheets. Open fours are usually easier for students to write. Your student should recognize fours and ones in both the handwritten and printed forms and know that both forms mean the same thing.

You can also use the first two houses in Decimal Street® (lesson 1) to teach counting to 20. Begin by placing one green unit block in the units house and saying and writing 1. Add another unit block and say and write 2. Keep doing this until there are 9 at home in the units house. Then try to add one more unit and notice that there is no more room. The reality at this point is that the 10 individual green units must be transformed into a blue ten bar to make room.

To make it more interesting, we can make up a story about these new units who keep moving into our neighborhood. Apparently they heard that we have a nice home, and naturally they want to live with us. After we welcome them into our home, we find that we like them as well. However, when the house is full (nine units) and a new fellow wants to live with us, we have a family discussion. We decide that in order for him to live with us, we will all have to become a blue ten bar and move into the house next door. That way we can all be together.

As more units arrive, they can stay in the units house because it is now open. When the first unit arrives after we have moved into the tens house, we have one ten bar and one unit bar, or 11. The next one arrives and there are 12—one ten and two units—living on our street. Eventually we have 19 units. When the next one arrives, another 10 is formed and moves in with the original 10, so now there are two tens, or 20. This is illustrated on the next page.

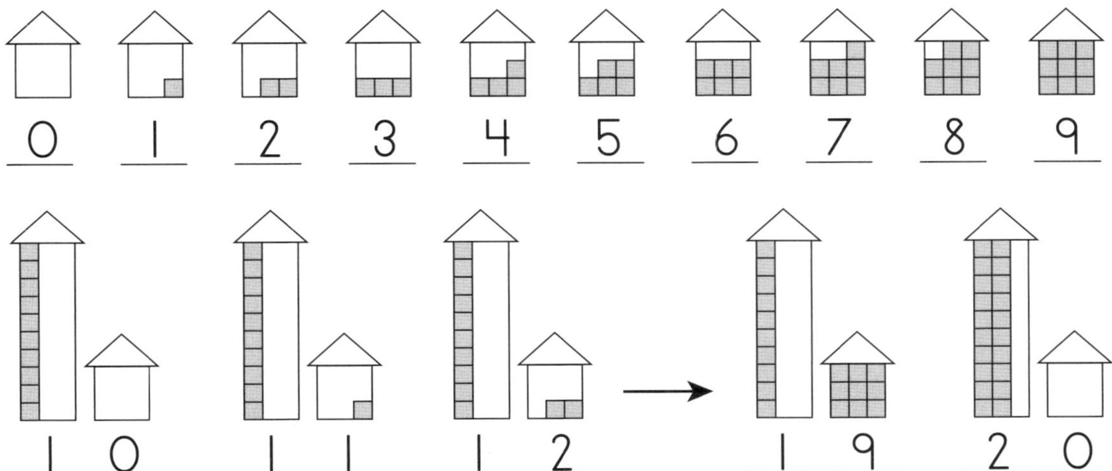

When counting to 20, encourage the student to count aloud using the common names: "one, two . . . eight, nine, ten, eleven, twelve . . . twenty," as well as with the place value names: "eight, nine, onety, onety-one, onety-two . . . two-ty." The common names are used most often, but the place value names can help the student understand the concept.

LESSON 3

Unit Bars

Now we will help the student transition from the green units. When teaching place value, I used the red, blue, and green blocks. Before attempting addition, be sure the student knows that a group of three green units is the same as one pink unit bar, which is three units long. Place the blocks side by side to teach this. Show that the three individual green units "glued together" are the same length as the three bar. This concept is very important, as it is the basis of addition—two units plus three units have the same value as five units. We want to transition the student from using the green units exclusively to using the colored unit bars as well. On the worksheet, match the units on the left with the correct bar on the right. You can have the student color the bars on the right to match the blocks.

There are four levels of combining or arithmetic:

1. Counting
2. Adding, which is fast counting
3. Multiplying, which is fast adding
4. Exponents, which are fast multiplying

To move from counting to adding, the student has to learn the values of the bars so that we can add two plus three and get five, instead of counting "one, two," and then "three, four, five."

Word		#		Unit pieces		Unit bar and color
two	=	2	=	□□	=	□□ orange
three	=	3	=	□□□	=	□□□ pink
four	=	4	=	□□□□	=	□□□□ yellow
five	=	5	=	□□□□□	=	□□□□□ light blue
six	=	6	=	□□□□□□	=	□□□□□□ violet or purple
seven	=	7	=	□□□□□□□	=	□□□□□□□ tan or vanilla
eight	=	8	=	□□□□□□□□	=	□□□□□□□□ brown
nine	=	9	=	□□□□□□□□□	=	□□□□□□□□□ light green

Measurement Using Objects

The language of measurement may be taught here as well. For example, use five unit blocks side by side to measure the length of the blue five bar. We can say that the five bar is five units long. Also, use the blocks to measure everyday objects. For example, ask the student to measure a pencil with the unit blocks or the top of a desk with the ten bar turned on its side. This is preparation for measurement with standard units, such as feet and inches (introduced in *Beta*).

Games for Unit Bar Identification

Simon Says - Tell the student, "Put a three block on your nose," or "Hide two five blocks in your pocket." Be as creative as you wish.

What's Missing? - Put the one through nine blocks on the table. Ask the student to cover his eyes while you remove one of the blocks and then ask him which one is missing. Take turns. There are numerous variations to this. Try removing two blocks, or you can start with two of each number and remove one or two blocks.

The Grab Bag - Put the one through nine blocks in an opaque bag. Take turns either drawing a number card and finding a particular block or telling each other to feel around and find a certain block. An easier version is to simply name the one you are about to pull out. A harder version is to name the missing block after it has been removed from the bag.

Blocks and Digits Match-Up - Make a set of 3 × 5 cards with the digits 1 through 9 on them (or use the cards from the application and enrichment pages in the student workbook). On the back of the cards draw colored dots to match the blocks. Place the cards digit side up and match the blocks to them. This is especially good as an activity that students can get out on their own, as it is self-correcting. Go ahead and let children peek; they stop soon enough.

Sing and Grab - "If you're happy and you know it" . . . "Clap three times," or "Grab a five," etc.

Act It Out - Let students choose a way to act out the number represented by a block. For example, for a three block, clap three times, jump three times, write three lines on a piece of paper, play three notes on the piano, or eat three raisins. Encourage students to use their imaginations for this!

Teaching Tip

As you get ready to begin teaching the addition facts, be sure to give the student time to "play" with the blocks. Some schools schedule a portion of each day for "play" and refer to this time as "free exploration."

As the student builds, he automatically learns the relationships between the different numbers. For example, if a six bar is laid on top of a ten bar, it doesn't take long for most students to realize that a four bar is needed to finish the row. When it is time to learn 6 + 4 = 10, the student already understands the reality that the fact describes.

LESSON 4

Addition: Symbols, +0

Word Problem Tips

As we begin addition and review place value, employ as many senses as possible. Use the blocks! Even though the picture is sufficient in this case, the more you use the blocks, the better will be the student's understanding. That is why we build first. Have the student color the worksheets to make the connection from the blocks to the picture. Then match with the correct answer on the right side of the worksheet. After this, write the correct numbers on the lines below the picture. Once the problem has been solved, read the equation out loud. The examples show how to do a problem while building, writing, and saying.

As a readiness exercise, practice counting up by ones from zero to nine and backwards from nine to zero. This is preparation for the one facts and the nine facts. Have the student build the steps with the unit bars as shown in Figure 1.

Figure 1

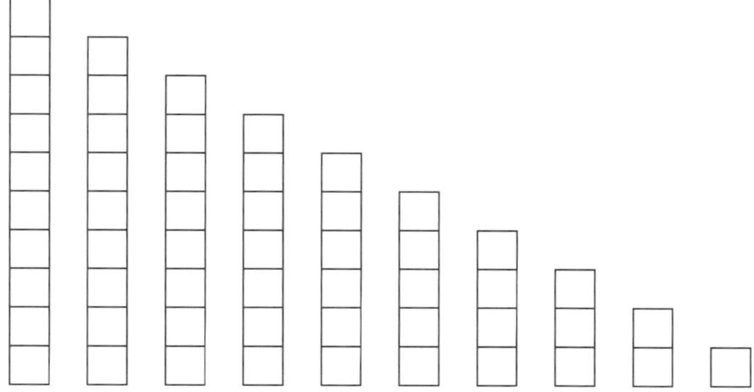

Students understand the concept of "nothing." When teaching 2 + 0, illustrate by saying something like, "We have two dogs, and we didn't get any more today. This means we have two plus no more, or 2 + 0 = 2."

Example 1

Solve: 2 + 0

Place another two bar and "nothing" above the first two bar.

orange	+		=	orange		Build.
2	+	0	=	2	2 + 0 = 2	Write.

"Two plus zero is the same length as two." Say.

Example 2

Solve: 0 + 3

Place another three bar and "nothing" above the first three bar.

	+	pink	=	pink		Build.
0	+	3	=	3	0 + 3 = 3	Write.

"Zero plus three is the same length as three." Say.

Ordering by Length

As students work with the blocks, use this opportunity to teach the vocabulary of comparison. For example, compare a three bar, a five bar, and a ten bar. The five bar is ***longer*** than the three bar, and the ten bar is the ***longest***. Point out that if the ten bar is longer than the five bar and if the five bar is longer than the three bar, then the ten bar must be longer than the three bar. Similarly, use ***short***, ***shorter***, and ***shortest*** to compare and describe three different lengths. Continue to use the language of comparison in everyday life to compare attributes. For example, compare the lengths of three pencils or the heights of three different people.

Addition Chart

We will be reinforcing the concept of the ***Commutative Property of Addition*** in later lessons. According to the Commutative Property, 0 + 2 has the same answer as 2 + 0, so if you know one fact, you automatically know the other. This is encouraging because it reduces the number of facts to be learned. This concept is difficult to illustrate visually for adding zero, so if you want to wait until a later lesson to mention this concept, feel free to do so.

The facts that are studied in each lesson will be shaded on the chart as we progress. There are 100 facts on the chart, but because of the Commutative Property, we really only need to learn 50 different facts. When a student has mastered those facts, have him circle, underline, or color the same facts on his own sheet (after lesson 4 in the student workbook).

0+0	0+1	0+2	0+3	0+4	0+5	0+6	0+7	0+8	0+9
1+0	1+1	1+2	1+3	1+4	1+5	1+6	1+7	1+8	1+9
2+0	2+1	2+2	2+3	2+4	2+5	2+6	2+7	2+8	2+9
3+0	3+1	3+2	3+3	3+4	3+5	3+6	3+7	3+8	3+9
4+0	4+1	4+2	4+3	4+4	4+5	4+6	4+7	4+8	4+9
5+0	5+1	5+2	5+3	5+4	5+5	5+6	5+7	5+8	5+9
6+0	6+1	6+2	6+3	6+4	6+5	6+6	6+7	6+8	6+9
7+0	7+1	7+2	7+3	7+4	7+5	7+6	7+7	7+8	7+9
8+0	8+1	8+2	8+3	8+4	8+5	8+6	8+7	8+8	8+9
9+0	9+1	9+2	9+3	9+4	9+5	9+6	9+7	9+8	9+9

Word problems begin in this lesson. Read them to the student, if necessary. You may use the blocks to illustrate the problems. On the next page are some general tips for teaching word problems.

Word Problem Tips

It is often challenging to teach children how to solve word problems. Here are some suggestions for helping your student learn this important skill. Don't be surprised if students need help with word problems for some time after they can do number problems involving the same concepts.

The first step is to realize that word problems require both reading and math comprehension. Don't expect a child to be able to solve a word problem if he does not thoroughly understand the math concepts involved. On the other hand, a student may have a math skill level that is stronger than his or her reading comprehension skills. Below are a number of strategies to improve comprehension skills in the context of story problems. You may decide which ones work best for you and your child.

Strategies for word problems:

1. Ignore numbers at first and read the story. It may help some students to read the question aloud. Every word problem tells a story. Before deciding what math operation is required, let the student retell the story in his own words. Who is involved? Are they receiving gifts, losing something, or dividing a treat?

2. Relate the story to real life, perhaps by using names of family members or friends. For some students, this makes the problem more interesting and relevant.

3. Build, draw, or act out the story. Use the blocks or actual objects when practical. Especially in the lower levels, you may require the student to use the blocks for word problems even when the facts have been learned. Don't be afraid to use a little drama as well. The purpose is to make it as real and meaningful as possible.

4. Look for the common language used in a particular kind of problem. Pay close attention to the word problems on the lesson practice pages, as they model different kinds of language that may be used for the new concept just studied. For example, "altogether" often indicates addition. These key words can be useful clues, but they should not be a substitute for understanding.

5. Look for practical applications that use the concept and ask questions in that context.

6. Have the student invent word problems to illustrate number problems from the lesson.

Cautions:

1. Unneeded information may be included in the problem. For example, we may be told that Suzie is eight years old, but the eight is irrelevant when adding up the number of gifts she received.

2. Some problems may require more than one step to solve. Model these questions carefully.

3. There may be more than one way to solve some problems. Experience will help the student choose the easier or preferred method.

4. Estimation is a valuable tool for checking an answer. If an answer is unreasonable, it is possible that the student used the wrong method to solve the problem.

Addition Facts Sheet

0 + 0	0 + 1	0 + 2	0 + 3	0 + 4	0 + 5	0 + 6	0 + 7	0 + 8	0 + 9
1 + 0	1 + 1	1 + 2	1 + 3	1 + 4	1 + 5	1 + 6	1 + 7	1 + 8	1 + 9
2 + 0	2 + 1	2 + 2	2 + 3	2 + 4	2 + 5	2 + 6	2 + 7	2 + 8	2 + 9
3 + 0	3 + 1	3 + 2	3 + 3	3 + 4	3 + 5	3 + 6	3 + 7	3 + 8	3 + 9
4 + 0	4 + 1	4 + 2	4 + 3	4 + 4	4 + 5	4 + 6	4 + 7	4 + 8	4 + 9
5 + 0	5 + 1	5 + 2	5 + 3	5 + 4	5 + 5	5 + 6	5 + 7	5 + 8	5 + 9
6 + 0	6 + 1	6 + 2	6 + 3	6 + 4	6 + 5	6 + 6	6 + 7	6 + 8	6 + 9
7 + 0	7 + 1	7 + 2	7 + 3	7 + 4	7 + 5	7 + 6	7 + 7	7 + 8	7 + 9
8 + 0	8 + 1	8 + 2	8 + 3	8 + 4	8 + 5	8 + 6	8 + 7	8 + 8	8 + 9
9 + 0	9 + 1	9 + 2	9 + 3	9 + 4	9 + 5	9 + 6	9 + 7	9 + 8	9 + 9

LESSON 5

Addition: +1, Commutative Property

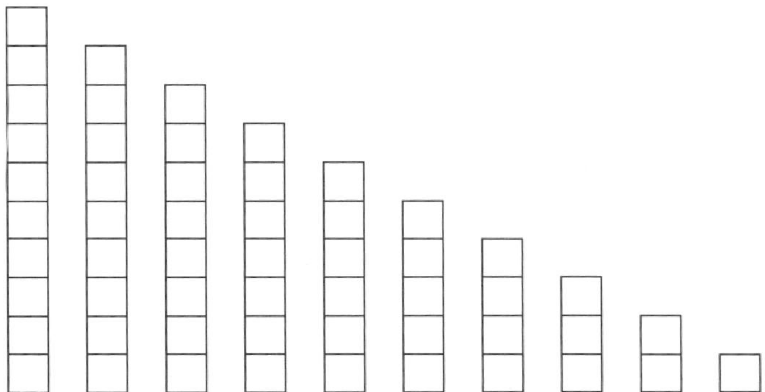

Here is a narrative to use and adapt until the concept of adding one is learned.

Have the student put his/her finger on the three bar. Say, "Point to the bar of the number that is one more than that." The student should point to the four bar and say four. Continue to ask questions: "What is one greater than that? . . . If you add one to five, what do you have? . . . What is one greater than six? . . . If I add one to seven what do I have? . . . What is one more than eight?"

Keep going until the student knows "one more than." Change your vocabulary in order to teach different important words that indicate addition.

When showing the one facts, begin by taking the green one (unit) bar and placing it end to end with the pink three bar. Ask the student if he can find another bar that is the same length as the one bar and the three bar "smooshed" together. This is addition—placing one bar end to end with another bar and finding a third bar that is the same length (in this case, the yellow four bar). Now write $1 + 3 = 4$ next to the bars while saying, "One plus three is the same length as four," or "One plus three equals four."

The child sees the problem, builds it, writes it, reads it, and hears it. After this repetition, he should remember it and understand it!

Example 1

Solve: 3 + 1

Place the three and the one above the four.

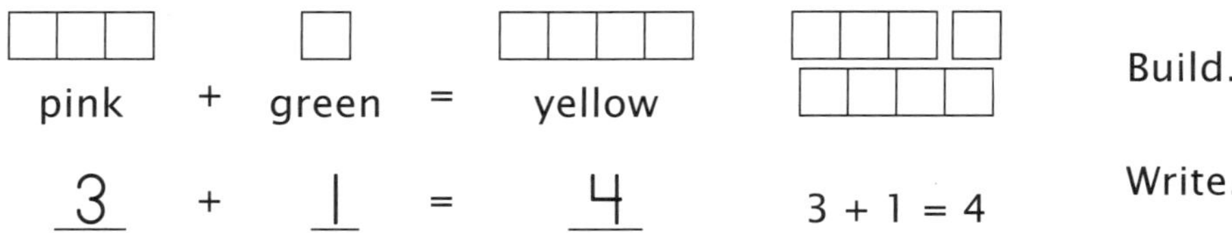

"Three plus one is the same length as, or equal to, four."

Say.

You can also show addition vertically as in example 2.

Example 2

Solve:
$$\begin{array}{r} 1 \\ +\ 5 \\ \hline \end{array}$$

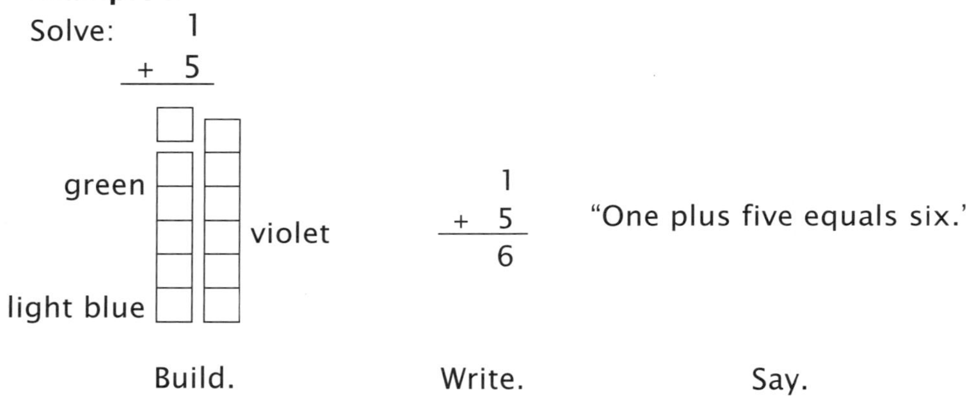

Now that we are using two blocks and "smooshing" them together to make a third number, introduce the Commutative Property of Addition. When introducing this term, relate it to a person who commutes to work. If it is 12 miles to work one way, then it is 12 miles back home. No matter which way he is traveling, he is still going 12 miles.

The ***Commutative Property of Addition*** means that we can change the order of an addition problem without changing the answer. In other words, 1 + 5 is the same as 5 + 1. This is an abstract concept that students have to see to understand, as they generally look at 1 + 5 and 5 + 1 as two different math facts.

Once you have taught the Commutative Property, show what a useful thing it is; every time we learn one math fact, we are really learning two on the chart. Having memorized the zero and one facts, we have learned 36 out of 100 facts. That is a good start!

0+0	0+1	0+2	0+3	0+4	0+5	0+6	0+7	0+8	0+9
1+0	1+1	1+2	1+3	1+4	1+5	1+6	1+7	1+8	1+9
2+0	2+1	2+2	2+3	2+4	2+5	2+6	2+7	2+8	2+9
3+0	3+1	3+2	3+3	3+4	3+5	3+6	3+7	3+8	3+9
4+0	4+1	4+2	4+3	4+4	4+5	4+6	4+7	4+8	4+9
5+0	5+1	5+2	5+3	5+4	5+5	5+6	5+7	5+8	5+9
6+0	6+1	6+2	6+3	6+4	6+5	6+6	6+7	6+8	6+9
7+0	7+1	7+2	7+3	7+4	7+5	7+6	7+7	7+8	7+9
8+0	8+1	8+2	8+3	8+4	8+5	8+6	8+7	8+8	8+9
9+0	9+1	9+2	9+3	9+4	9+5	9+6	9+7	9+8	9+9

Teaching Tip

Using a dry erase board can be helpful for students who have difficulty with fine motor skills. After you have practiced with the blocks, have the student work problems from the lesson on the board. This can also be helpful for students who need a chance to move around while learning. Take this one step further by going outside and using sidewalk chalk!

LESSON 6

Counting to 100, Skip Counting by 10

Counting Past 100

Once a student can count to 20, the progression to counting to 100 should happen naturally. Notice that "two-ty" is twenty, "three-ty" is thirty, "four-ty" is forty, "five-ty" is fifty. Sixty, seventy, eighty, and ninety are all consistent with the pattern of saying the unit name and adding the suffix *-ty* for ten, so there is no need to differentiate between the common name and the place value name.

Have the student practice filling in the numbers on the blank lines on the student pages. Require this only enough for the student to demonstrate the ability to do so. Practice orally by giving any number between one and 100 and having the student count from the given number to 100.

Skip Counting by 10

There are several reasons for teaching skip counting. I will mention one reason now and then more in a later lesson.

Skip counting teaches the concept of multiplication, which is fast adding of the same number. In learning the tens, we are able to count groups of ten quickly. Notice how it reinforces writing to 100. The numbers in the far left column of the hundred chart are the multiples of 10: 10, 20, 30, and so on.

A good way to practice skip counting is to read the numbers softly from zero to nine and then have the student shout "10!" Continue counting softly—11, 12, 13 . . . 18, 19—and have the student shout "20!", repeating until the student can skip count from 10 to 100.

Some practical examples of counting by 10 are fingers on both hands, toes on both feet, and pennies in a dime. Notice that the multiples of 10—20, 30, 40, etc.—all end in *-ty*. The suffix *-ty* represents ten. Thus, six-ty means six tens, and seven-ty means seven tens.

Example 1

Skip count and write the numbers on the lines. Say them out loud as you count and write. Then write the numbers in the spaces provided beneath the figure.

									10
									20
									30
									40
									50
									60

____, ____, 30, 40, ____, ____, ____

Example 2

Fill in the missing information on the lines.

____, ____, ____, 40, ____, ____, 70, 80, 90, ____

Solution

10, 20, 30, 40, 50, 60, 70, 80, 90, 100

Counting Past 100

When students have mastered counting to 100, encourage them to continue counting past 100 to 120 or further. Use the blocks and their knowledge of place value as needed. Say each number without saying "and" between the hundreds place and the tens or units places. In other words, do not say "one hundred and one" or "one hundred and fifty."

Example 3

One hundred, one hundred one, one hundred two, . . . one hundred ten, one hundred eleven, . . . one hundred twenty

LESSON 7

Addition: +2

Place Value

Before you begin teaching the two facts, practice counting up by twos to ten: 0-2-4-6-8-10. Use the blocks arranged like steps as we did with counting by ones. These are the ***even numbers***. Then practice the same skill beginning with one: 1-3-5-7-9. These numbers are the ***odd numbers***. Going up by two reinforces the two facts. See the game for pre-addition of twos at the bottom of the page.

Example 1

Solve: 5 + 2

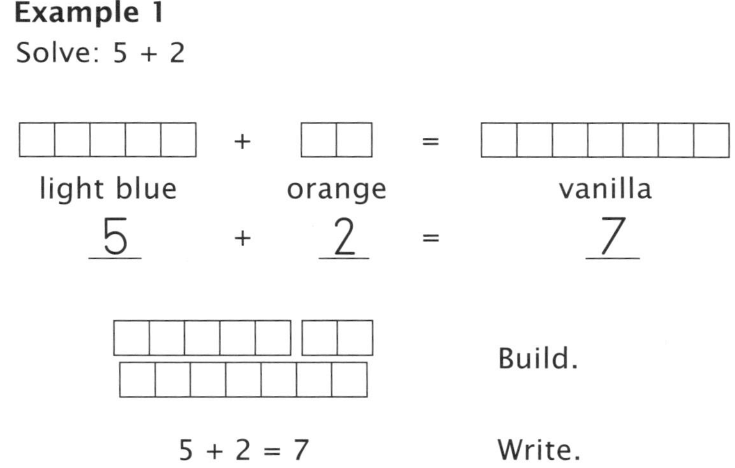

Five plus two is equal to seven. Say.

Game for Pre-Addition of Twos

Bigger – Get out the one through nine blocks and stack them in ascending order starting on the right with the green unit. Directly to the left of the unit is the orange two bar, then the pink three bar, and on up to the nine bar. This is identical to Figure 1 in lesson 4. Have the child touch the yellow four bar and then ask the question, "What number is two more than four?" Then have the child touch, or point to, the violet six bar and say "six." Do this for all the unit bars. When this is mastered, simply ask what number is two more than each unit bar.

Example 2

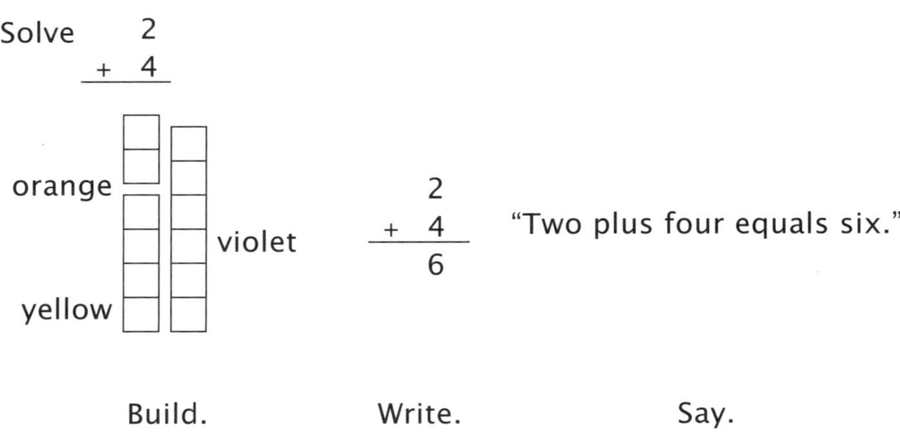

Build. Write. Say.

In the addition problem in Example 1, the numbers 5 and 2 are referred to as the ***addends***, and 7 is the ***sum***. In Example 2, the numbers 2 and 4 are the addends, and 6 is the sum. We add the two addends to form the sum in an addition problem. Use the terms often so that the student becomes familiar with them. For example, when discussing the Commutative Property, state that we can change the order of the addends without changing the sum.

Place Value

To keep our place value skills fresh and teach a foundational concept of mathematics, we are going to add tens and hundreds as well as units. When teaching adding, subtracting, and comparing (less than or greater than), I often say, "To compare or combine you must be the same kind." Numbers to be added must have the same place value. When subtracting, the numbers must be the same kind as well. We can't add apples to oranges. We can only add apples to apples and oranges to oranges.

So far our adding has been units to units. In Example 2 we showed that two units plus four units is the same as six units. In Example 3 we show that one hundred plus one hundred is two hundreds.

Example 3

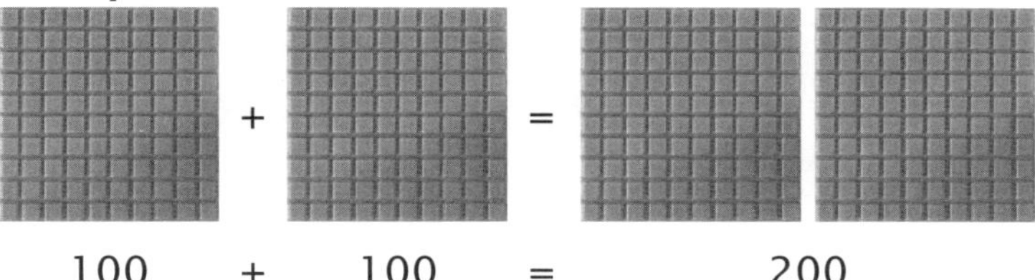

100 + 100 = 200

In Example 4 we see that four tens plus two tens is the same as six tens.

Example 4

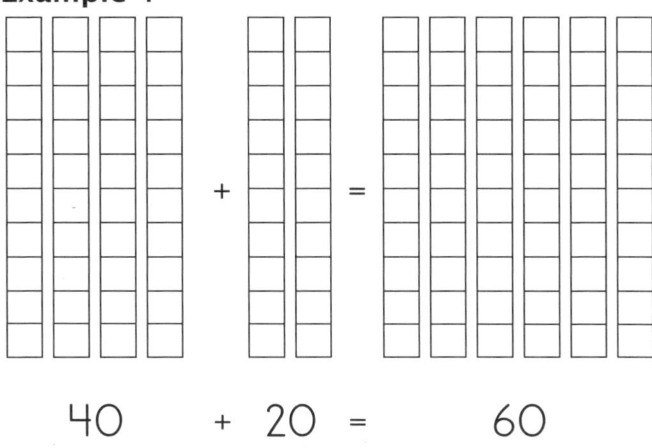

40 + 20 = 60

Because of the Commutative Property, we see that after learning the two facts, we have learned 51 out of 100 facts. That is over half!

0 + 0	0 + 1	0 + 2	0 + 3	0 + 4	0 + 5	0 + 6	0 + 7	0 + 8	0 + 9
1 + 0	1 + 1	1 + 2	1 + 3	1 + 4	1 + 5	1 + 6	1 + 7	1 + 8	1 + 9
2 + 0	2 + 1	2 + 2	2 + 3	2 + 4	2 + 5	2 + 6	2 + 7	2 + 8	2 + 9
3 + 0	3 + 1	3 + 2	3 + 3	3 + 4	3 + 5	3 + 6	3 + 7	3 + 8	3 + 9
4 + 0	4 + 1	4 + 2	4 + 3	4 + 4	4 + 5	4 + 6	4 + 7	4 + 8	4 + 9
5 + 0	5 + 1	5 + 2	5 + 3	5 + 4	5 + 5	5 + 6	5 + 7	5 + 8	5 + 9
6 + 0	6 + 1	6 + 2	6 + 3	6 + 4	6 + 5	6 + 6	6 + 7	6 + 8	6 + 9
7 + 0	7 + 1	7 + 2	7 + 3	7 + 4	7 + 5	7 + 6	7 + 7	7 + 8	7 + 9
8 + 0	8 + 1	8 + 2	8 + 3	8 + 4	8 + 5	8 + 6	8 + 7	8 + 8	8 + 9
9 + 0	9 + 1	9 + 2	9 + 3	9 + 4	9 + 5	9 + 6	9 + 7	9 + 8	9 + 9

We will be reviewing addition and subtraction facts throughout the student workbook. If you find that you need more review of these facts, visit mathusee.com for more resources.

Teaching Tip

Some students respond well to timed practice and any kind of competition. Others find timed drills very intimidating. Be sure to adapt the drills to fit your student. For a non-threatening way to practice, try having the student quiz a parent or teacher on the math facts. After filling in a practice page and making sure the answers are correct, the student reads the problem to the teacher. If the teacher seems confused or answers incorrectly, the child can correct the answer from the worksheet.

This is a good time to introduce the words ***true*** and ***false***. Encourage the student to use these words when responding to the answers given by the teacher or parent. Teaching is a powerful way of learning, so put it to work for you.

LESSON 8

Solve for an Unknown

I like to teach solving for an unknown at the same time I am teaching the addition facts. There are three very important reasons for this.

1. It reinforces the basic facts.

2. It provides a foundation for subtraction.

3. It familiarizes the students with algebra.

Let's work through some examples. Notice that we are not teaching algebra abstractly (just letters and numbers on paper) but concretely with the manipulatives, in order to give meaning to the letters and numbers. **When solving for the unknown, allow the student to use the blocks for as long as they are needed.**

Example 1

1. Say, "What number plus two is the same as nine?" ? + 2 = 9

2. Build. Put the two above the nine and find the missing piece.

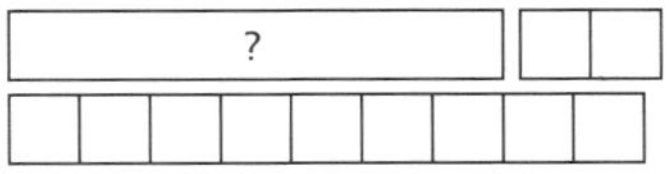

3. Write 7 in the space with the question mark. 7 + 2 = 9

4. Say, "Seven plus two is the same as nine."

Verbalizing this correctly is essential to the student's understanding. When you are looking for the correct unit bar, don't be afraid to experiment. I usually reach for the four bar or the five bar first and try it. I want to show that it is okay to eliminate possibilities, and I like to encourage a student's attempts to experiment to find the correct solution. Later we can move to writing X + 2 = 9. We use a letter because we don't know exactly what number makes this the same, or equal, in an equation. I tell students, "When we don't know what number to use, pick a letter!"

Although we are happy that the student can solve for an unknown, we are not satisfied until he can make a word problem out of this equation. After all, the end product of our math instruction is to apply it to real-life situations. For Example 1, you might say that you need nine dollars by the weekend, and you have two dollars. How much more do you need? As the student does the worksheets, encourage him or her to make word problems to fit the equations. This will help later when the student is asked to write an equation for the word problem.

Remember to continue to practice counting from 0 to 100.

Example 2

1. Say, "What number plus two is the same as five?" _?_ + 2 = 5
2. Place the two above the five and find the missing piece.

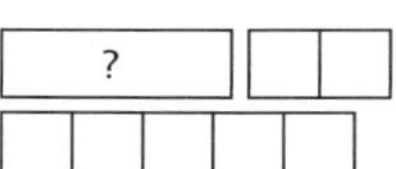

3. Write 3 in the space with the question mark. _3_ + 2 = 5
4. Say, "Three plus two is the same as five."

Games for Problem Solving

Who Are You? Who Am I? – Example: "Together we are seven. You are a one; who am I?" When the student has mastered numbers added up to 10, extend the numbers past 10.

Both Sides the Same – Get a large piece of white paper and draw a line down the middle. Place a number bar on one side and a smaller bar on the other side, with a third piece hidden under a bowl or napkin. For example, place a seven bar on one side; showing on the opposite side is a two bar and a bowl. Ask what is hidden under the bowl and say, "What plus two is the same as seven?" When the student figures it out, pick up the bowl to reveal the five bar. Start with lesser numbers and place progressively greater numbers on the known side.

LESSON 9

Addition: +9, Mental Math

In this lesson we will be adding by nines. The idea of making or wanting to be ten will be your "fun-dation" for regrouping. First practice counting backwards by one, using the game at the end of this lesson. Taking one away, or counting down by one, is essential to our approach to learning to add by nine.

I like to introduce this with a short narrative about how nine isn't content because he wants to be ten. Ask most nine-year-olds how old they want to be, and they say, "Ten!" Children understand Mr. Nine. Next ask, "What does nine need to have added to him to be ten?" "One unit!" Nine is therefore always on the prowl, looking for one more so he can be ten. Using a nine bar and several green unit bars, create the equation 9 + 5. Be as dramatic as you like, perhaps having the student look away or close his eyes. In that instant, nine takes one to become ten (or "onety").

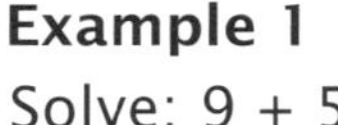

Example 1

Solve: 9 + 5

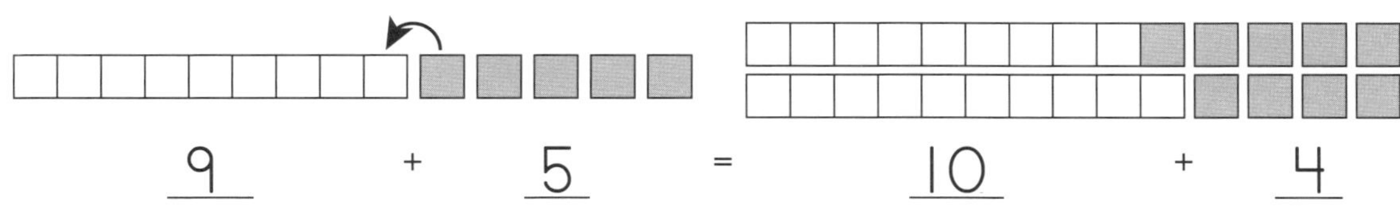

Nine plus five is equal to ten plus four, or fourteen.

Note: This will be the first time a student has added ten to a number. Simply apply what the student knows about place value. Start with ten and ask what you would have if you added two more. For example: 10 + 2 = 12. Put together a ten bar and a two bar to illustrate this.

In Example 1, we still have one nine and five units, which are the same length as one ten and four units. Nine is finally happy, and 10 + 4 is 14 ("onety-four"). We can also see that 9 + 5 = 14. The original five has been decreased by one from five to four, and nine has been increased by one to be ten. This is what regrouping is all about!

To provide a visual reminder of this, you can make the circle on the top of the numeral 9 the end of a vacuum nozzle. Nine is always "sucking up" one. Making the vacuum noise is fun and multi-sensory. When a child sees 9, she thinks "one less" and makes the appropriate sound. Practice the nines now until the student understands and feels confident adding by nine. Be sure to practice "taking one away" first with the game on the next page.

Another way to solve adding by nine is to use the colored unit bars. For 9 + 5, pick out the lime green nine bar and the light blue five bar. Place them end to end and say, "Nine plus five is the same as ten plus what?" Have the student find the yellow four bar and place it at the end of the blue ten bar. Then say, "Nine plus five is the same as ten plus four, or fourteen." See Example 2. Choose the way that helps the student understand the concept most effectively. Don't forget to use the same strategies as in previous lessons. Present the problems by building, writing, and saying to assist in memorizing and understanding these facts. You may also find the addition fact songs on the *Skip Counting CD* useful.

Example 2

Solve: 9 + 5

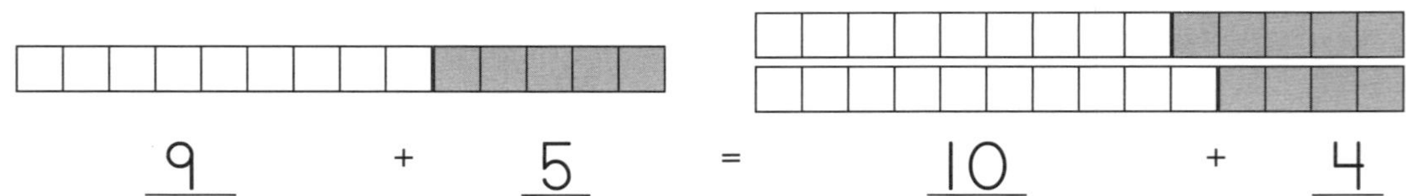

Nine plus five is equal to ten plus four, or fourteen.

With this lesson, we have learned 64 out of 100 facts. That is over half!

0+0	0+1	0+2	0+3	0+4	0+5	0+6	0+7	0+8	0+9
1+0	1+1	1+2	1+3	1+4	1+5	1+6	1+7	1+8	1+9
2+0	2+1	2+2	2+3	2+4	2+5	2+6	2+7	2+8	2+9
3+0	3+1	3+2	3+3	3+4	3+5	3+6	3+7	3+8	3+9
4+0	4+1	4+2	4+3	4+4	4+5	4+6	4+7	4+8	4+9
5+0	5+1	5+2	5+3	5+4	5+5	5+6	5+7	5+8	5+9
6+0	6+1	6+2	6+3	6+4	6+5	6+6	6+7	6+8	6+9
7+0	7+1	7+2	7+3	7+4	7+5	7+6	7+7	7+8	7+9
8+0	8+1	8+2	8+3	8+4	8+5	8+6	8+7	8+8	8+9
9+0	9+1	9+2	9+3	9+4	9+5	9+6	9+7	9+8	9+9

Game to Precede Adding by 9

Smaller – Get out the one through nine blocks and stack them in ascending order with the green unit on the right. Ask the question, "Which block is one unit smaller than the (   ) block?" or "What number is a one less than (   )?" Do this until the student knows each answer; only then move to learning the nine facts.

Mental Math

Mental math problems may be used to keep the facts alive in the memory and to develop mental math skills. The teacher should say the problem slowly enough so that the student comprehends it and then walk him through increasingly-difficult exercises. The purpose is to stretch but not discourage. See the example below, along with some suggested problems to try.

Example 3

2 + 3 + 1 = ? "Two plus three plus one equals what number?"

The student thinks, "2 + 3 = 5, and 5 + 1 = 6." At first you will need to go slowly enough for him or her to verbalize the intermediate step. As skills increase, the student should be able to give just the answer.

Starting with this lesson, every third lesson in the *Alpha* instruction manual will have some suggested mental math problems for you to read aloud to your student. Try a few at a time and remember to go quite slowly at first.

1. Four plus one plus one equals what number? (6)
2. Two plus two plus zero equals what number? (4)
3. Five plus one plus two equals what number? (8)
4. Three plus two plus two equals what number? (7)
5. Eight plus one plus five equals what number? (14)
6. One plus three plus zero equals what number? (4)
7. Six plus two plus one equals what number? (9)
8. Five plus two plus two equals what number? (9)
9. Seven plus two plus eight equals what number? (17)
10. Nine plus zero plus one equals what number? (10)

LESSON 10

Addition: +8

The "solve for the unknown" problems on the systematic review pages introduce the use of X for the unknown. When we don't know what number to use, we can use a letter (any letter may be used). Write the correct answer over or under the letter. The student doesn't need to write X = 3 at this point.

Before you begin addition by eight, practice counting backwards by twos. The student should be able to say, "8-6-4-2-0" and "9-7-5-3-1" before going any further. See the game near the end of this lesson.

Remember how we added nines by "vacuuming" one? Now we are going to vacuum two. This is adding by eight. Eight has two circles in the number, or two vacuum nozzles. Mr. Eight wants to be ten, just as Mr. Nine does. When we add eight to any number, we take two from that number and add a ten.

In Example 1, take the brown eight bar and place five green unit bars beside it. Make the vacuum noise and take two of the green unit pieces and place them next to eight, making ten. Put a ten bar on top of or beside the 8 + 2. The student should see that 8 + 5 is the same as 10 + 3, or 13.

Example 1

Solve: 8 + 5

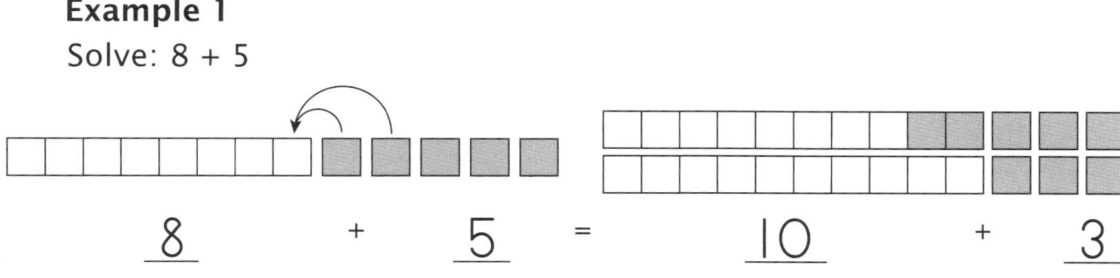

Eight plus five is equal to ten plus three, or thirteen.

We can solve adding by eight another way using the colored unit bars instead of the individual unit pieces.

Example 2
Solve: 8 + 5

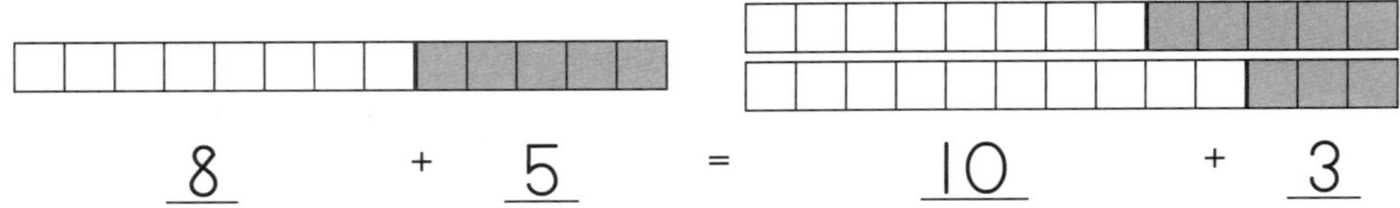

Eight plus five is equal to ten plus three, or thirteen.

Example 3
Solve: 8 + 7

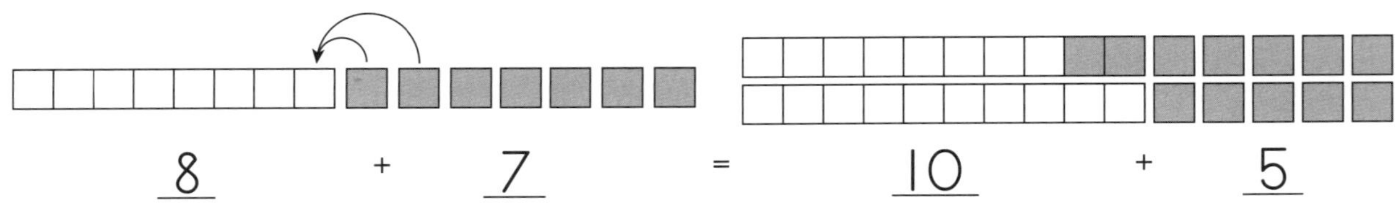

Eight plus seven is equal to ten plus five, or fifteen.

Example 4
Solve: 8 + 7

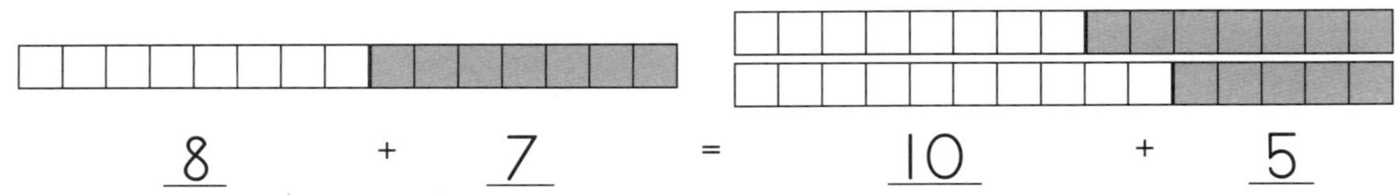

Eight plus seven is equal to ten plus five, or fifteen.

With the eight facts mastered, we now know 75 out of 100 facts. Good job! There are only 25 more to go.

0+0	0+1	0+2	0+3	0+4	0+5	0+6	0+7	0+8	0+9
1+0	1+1	1+2	1+3	1+4	1+5	1+6	1+7	1+8	1+9
2+0	2+1	2+2	2+3	2+4	2+5	2+6	2+7	2+8	2+9
3+0	3+1	3+2	3+3	3+4	3+5	3+6	3+7	3+8	3+9
4+0	4+1	4+2	4+3	4+4	4+5	4+6	4+7	4+8	4+9
5+0	5+1	5+2	5+3	5+4	5+5	5+6	5+7	5+8	5+9
6+0	6+1	6+2	6+3	6+4	6+5	6+6	6+7	6+8	6+9
7+0	7+1	7+2	7+3	7+4	7+5	7+6	7+7	7+8	7+9
8+0	8+1	8+2	8+3	8+4	8+5	8+6	8+7	8+8	8+9
9+0	9+1	9+2	9+3	9+4	9+5	9+6	9+7	9+8	9+9

Game to Precede Adding by 8

Smaller – Get out the one through nine blocks and stack them in ascending order with the green unit on the right. Ask the question, "Which block is two units smaller than the (   ) block?" or "What number is a two less than (   )?" Do this until they know each answer and then move to learning the eight facts.

Teaching Tip

Addition is the process of putting two numbers together to find a sum. It is also important for students to be able to take a number apart and find pairs of numbers that will add to the given number. The process is called ***decomposition***, but you do not need to use this word with your student.

The skill of decomposition should be a natural result of using the blocks as suggested throughout the Math-U-See program. You can teach it explicitly by showing children a number and asking for two or more numbers that will add to make the number. Have them show the answer with the blocks and write it using an equation—for example, 3 + 2 = 5.

Do this orally by saying the answer and letting the student give an acceptable problem to match that answer. Don't limit the response to facts that you have already studied. This is a good activity to do while traveling in the car.

Shapes: Circles and Triangles

Skip Counting by 2

A ***circle*** is a line drawn around a point to make a closed figure. Every point on the line is the same distance from the center. You can draw a circle with a compass. The distance between the point of the compass and the pencil is the same as the distance between the center and the circle. Another way to draw a circle is to tie one end of a piece of string to a nail and the other end to a pencil. Hold the nail, or nail it into a board. Keep the string taut and draw a circle with the pencil.

Look around you to find circles. Here are some examples: plates, bottoms of glasses and bottles, coins, clock faces, door knobs, ceiling lights, and tires.

A ***triangle*** is a closed figure with three straight sides and three points. Students should understand that triangles may look different from one another, but the defining characteristic is the three straight sides. The prefix ***tri-*** represents three. A tricycle has three wheels, and a tripod has three legs. Look around for examples of triangles in your environment.

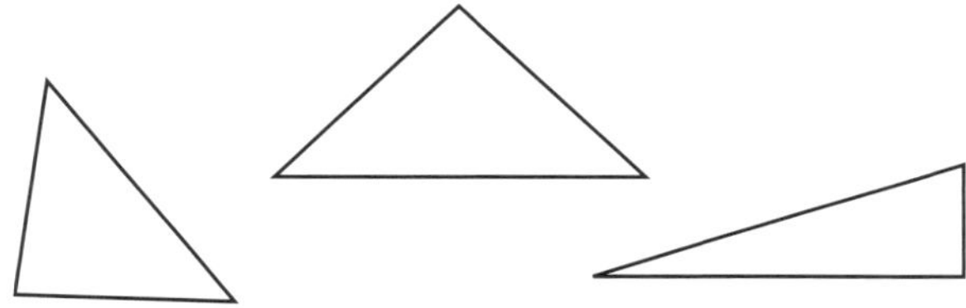

Skip Counting by 2

If you have previously completed the *Primer* curriculum, this will be review. Skip counting is a good foundation for many skills to be learned in the future.

Skip counting is the ability to count groups of the same number quickly. For example, if you were to skip count by threes, you would skip the one and the two and say "three;" then you would skip the four and the five and say "six," followed by "9-12-15-18." Skip counting by sevens is 7-14-21-28-35-42-49-56-63-70.

Here are four reasons for teaching skip counting.

1. It lays a solid foundation for learning the multiplication facts. The problem $3 + 3 + 3 + 3$ may be written as 3×4. If a child can skip count, he can say 3-6-9-12. Then he could read 3×4 as "3 counted 4 times is 12." As you learn your skip counting facts, you are learning all your multiplication facts in order.

2. Skip counting teaches the concept of multiplication. I had a teacher tell me that her students had successfully memorized their facts but didn't understand the concept. After she had taught them skip counting, they comprehended what they had learned. Multiplication is fast adding of the same number.

3. Skip counting is helpful as a skill in itself. It is a way of counting multiples of a certain number. A pharmacist attending a workshop told me he skip counts pills as he puts them into bottles.

4. Skip counting teaches you the multiples of a number that are needed when making equivalent fractions and finding common denominators. For example, $2/5 = 4/10 = 6/15 = 8/20$. The numbers 2-4-6-8 are the multiples of two, and 5-10-15-20 are the multiples of five.

In this lesson we will be counting by two. When introducing this, try pointing to each square. (See the examples on the next page.) As you count the squares, say the first number quietly and then ask the student to say the second number more loudly. Continue this practice, speaking more quietly each time until you are silently pointing to the first square while the student says the number loudly as you point to the second square. On the practice sheets, write numbers only in the squares with lines in them. See examples 1 and 2. In example 3, we are counting geometric shapes in groups of two.

Some practical examples of counting by two are counting eyes, ears, hands, feet, shoes, and socks. You might ask the student to count all the eyes or shoes in the family or in the classroom. This is also a good time to learn common prefixes like ***bi-***, which denotes two, as in bicycle or biped.

One parent asks his child to use his "inside voice" to verbalize the first number and his "outside voice" to say the second number. Another father with a military background has his son stand at attention and bark the skip count facts loudly in cadence.

Special Note to Teachers

In the *Alpha* curriculum we are going to learn to skip count by 2, 5, and 10. The rest of the skip count facts will be learned in *Gamma.* If you think that the student is ready to learn all of the skip counting facts now, feel free to go ahead. The *Skip Count and Addition Facts Song Book* and included CD can help children to learn these facts quickly. Have fun with the songs and facts.

Example 1

How many boxes are shown?

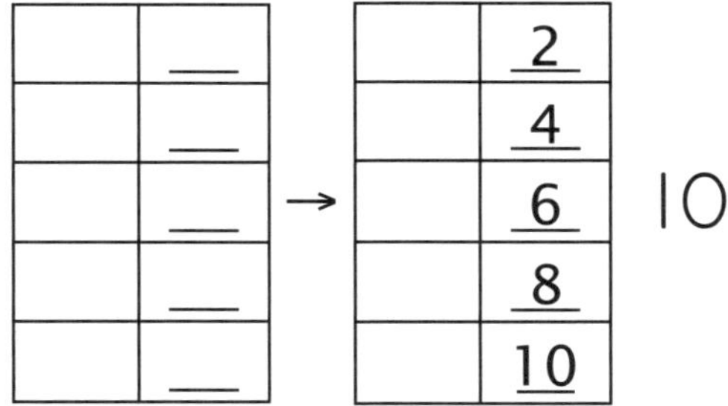

Example 2

How many boxes are shown?

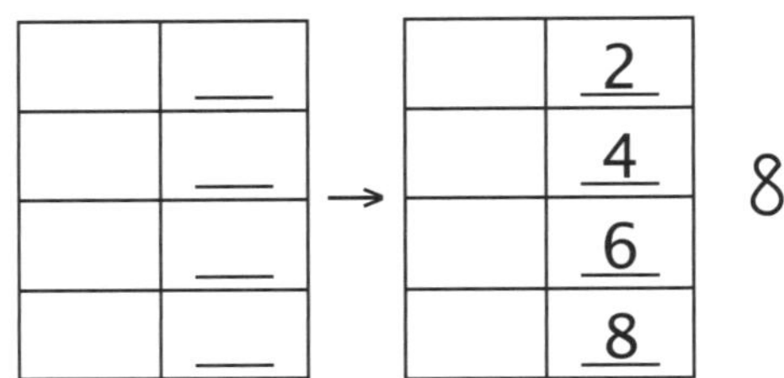

Example 3

How many triangles are shown?

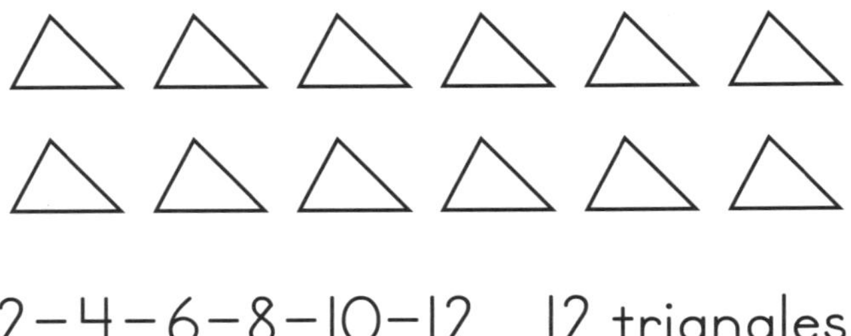

2–4–6–8–10–12   12 triangles

LESSON 12

Addition: Doubles

The next step in learning to add is to learn the doubles: 3 + 3, 4 + 4, 5 + 5, 6 + 6, and 7 + 7. Most children know these facts, except for 7 + 7. Here is 7 + 7 = 14 shown with the blocks.

Example 1
Solve: 7 + 7

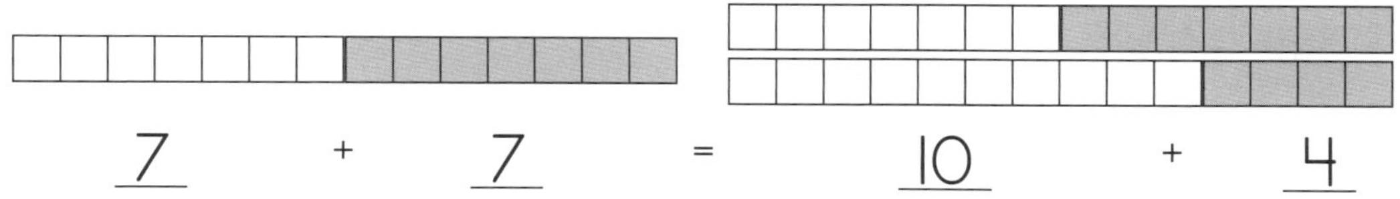

7 + 7 = 10 + 4

Seven plus seven is equal to ten plus four, or fourteen.

Example 2
Solve: 6 + 6

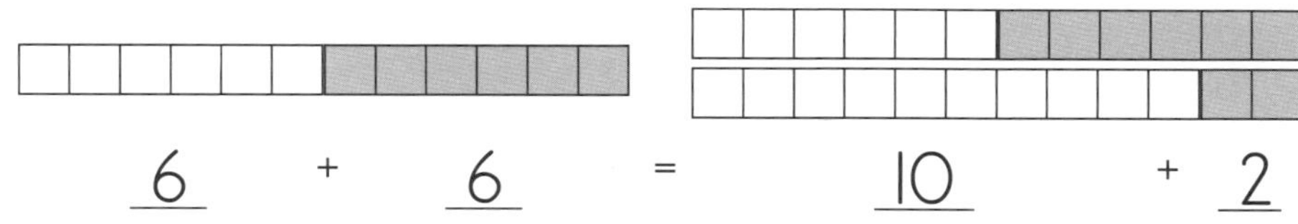

6 + 6 = 10 + 2

Six plus six is equal to ten plus two, or twelve.

Example 3
Solve: 5 + 5

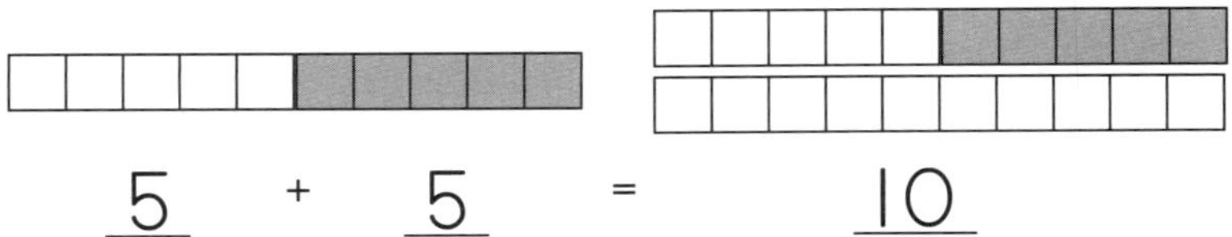

Five plus five is equal to ten.

At this stage younger children may or may not know how to read. As the teacher, use your discretion about how much to emphasize learning the number words. Here are the number words from zero to ten. On the student worksheets, the words will be provided so the student can select the appropriate answer.

0 - zero	4 - four	8 - eight
1 - one	5 - five	9 - nine
2 - two	6 - six	10 - ten
3 - three	7 - seven	

Example 4
Four plus two equals ________ .

4 + 2 = ___

Solution
Four plus two equals six.

4 + 2 = 6

With the addition of the doubles, we have finished memorizing 80 out of 100 facts. With only 20 facts left, you are to be congratulated.

0+0	0+1	0+2	0+3	0+4	0+5	0+6	0+7	0+8	0+9
1+0	1+1	1+2	1+3	1+4	1+5	1+6	1+7	1+8	1+9
2+0	2+1	2+2	2+3	2+4	2+5	2+6	2+7	2+8	2+9
3+0	3+1	3+2	3+3	3+4	3+5	3+6	3+7	3+8	3+9
4+0	4+1	4+2	4+3	4+4	4+5	4+6	4+7	4+8	4+9
5+0	5+1	5+2	5+3	5+4	5+5	5+6	5+7	5+8	5+9
6+0	6+1	6+2	6+3	6+4	6+5	6+6	6+7	6+8	6+9
7+0	7+1	7+2	7+3	7+4	7+5	7+6	7+7	7+8	7+9
8+0	8+1	8+2	8+3	8+4	8+5	8+6	8+7	8+8	8+9
9+0	9+1	9+2	9+3	9+4	9+5	9+6	9+7	9+8	9+9

Mental Math

Here are some more questions to read to your student.

1. Four plus four plus one equals what number? (9)
2. Two plus three plus five equals what number? (10)
3. One plus six plus two equals what number? (9)
4. Six plus two plus seven equals what number? (15)
5. Zero plus nine plus nine equals what number? (18)
6. One plus two plus three equals what number? (6)
7. Two plus four plus eight equals what number? (14)
8. Eight plus zero plus four equals what number? (12)
9. Three plus three plus nine equals what number? (15)
10. Two plus two plus two equals what number? (6)

Teaching Tip 1

With most children I have worked with, once you show them a few examples, they are ready to try one themselves. Watching the video and doing a few examples is what I call a normal presentation of a new concept, but with one particular student I tutored, that didn't work. What I did instead was to present the material and then do four or five examples without asking him to participate—just to watch. I would take a problem, build it, write it, and talk through the process as I worked toward the solution. I did this for all the problems. Sometimes I would ask him to select the problems that I should solve. That was all I did for the first day of a new lesson. The second day I would do a few more problems and then ask him to do one or two. By that time he had seen me do at least six problems before he felt comfortable enough to try one himself.

The reason I came upon this approach was that I noticed two examples were not nearly enough. He needed more time to assimilate a new concept. Because of his inability to grasp the new material in a short amount of time, he would become anxious and be even less able to understand the topic at hand. When he knew that he would not be required to participate the first day, he was able to relax and learn without becoming nervous about the prospect of having to do a problem after only a few examples.

I also realized how effective correct input and modeling is to learning. By taking the time to present the material clearly and correctly, the input improved dramatically, and the student was able to comprehend better because he knew exactly what was expected.

This student also had short-term memory issues. I couldn't spend too many days on new material, or he would forget what we learned in the previous lesson. I had to move to the review sheets after just a few days with the new material and intersperse the lesson practice sheets with the systematic review sheets.

Teaching Tip 2

Now that the student has learned a few sets of math facts, here is a tip to help in committing the facts to memory. When solving for an unknown, one addend and the sum are given, and we have to find the missing addend. In $X + 2 = 7$, we know the 2 and the 7, and we have to find the missing 5. The answer is $5 + 2 = 7$.

Solving for the unknown reviews addition, takes beginning steps in algebra, and lays a foundation for subtraction. In subtraction we will have $7 - 2 = 5$. These are the same numbers as $5 + 2 = 7$, just in a different order. The numbers 2, 5, and 7 can be called a trio. You have already covered adding by 8 and by 9. To cement these facts in your students' memories, come at what they have learned from a slightly different angle by using trios. Given 4 and 13, ask your student to find the missing number in the trio. The answer is 9. Explain that this technique is very similar to how we solved for an unknown in such problems as X + 4 = 13 or 4 + X = 13. You are simply being asked to find the missing information in a different way.

I hope these tips have been helpful.

Steve Demme

LESSON 13

Rectangles, Squares; Skip Counting by 5

The word ***rectangle*** means "right angle." It comes from the Latin word "rectus," which means "right." German, French, Spanish, and other languages have similar words with the same meaning. A ***right angle*** is a square corner. A closed shape with four square corners is a rectangle. Look around you for rectangles, such as this piece of paper. See how many other examples the student can identify.

Notice that the opposite sides of a rectangle have the same length.

A ***square*** is a special kind of rectangle. It has four right angles, so it is a rectangle. However, we also notice that all four sides have the same length. A rectangle with all four sides the same length is called a square. Here's another way of thinking of it: a square has four right angles and four sides that have the same length.

Skip Counting by 5

In this lesson we are reviewing skip counting by five. Use the same techniques that you used for skip counting by two and ten to introduce and teach this important skill. Some practical examples are fingers on one hand, toes on a foot, pennies in a nickel, players on a basketball team, and sides of a pentagon.

Example 1

Skip count and write the numbers on the lines. Say them out loud as you count and write. Then write them in the spaces provided beneath the figure.

				5
				10
				15
				20

5, ____, ____, ____

Example 2

Fill in the missing numbers on the lines.

5, ____, 15, ____, ____, 30, ____, 40, 45, ____

Solution

5, 10, 15, 20, 25, 30, 35, 40, 45, 50

Telling Time

Now that your student can skip count by five, he or she has the necessary skills to learn how to tell time. Telling time was presented in *Primer* and is taught again in *Beta,* the next book in the Math-U-See sequence. If you want to review what you learned in *Primer* or to introduce telling time now, consult Appendices A and B following lesson 30 in this book.

Teaching Tip

After learning the names of the basic shapes, it is helpful for students to have experience in manipulating the shapes and seeing how they are related to one another. Using the Math-U-See blocks provides some of this practice. For more experience working with different two-dimensional shapes, you can use attribute blocks, tangrams, or household objects. Attribute blocks may be used to sort by color, shape, and size. Tangrams provide challenging practice in putting together different geometric shapes to make new designs.

Students should be able to distinguish between flat (two-dimensional) and solid (three-dimensional) objects. At this level, students do not know the words two-dimensional and three-dimensional, but they should learn the names for the three-dimensional shapes.

Look for examples of different three-dimensional shapes in your everyday environment, just as you did for two-dimensional shapes. For example, use a box for a cube or rectangular prism, a can from the pantry for a cylinder, and an ice cream cone or party hat for a cone. If you have a block set that includes cubes, rectangular prisms, cylinders, or cones, encourage students to put these shapes together to make new shapes.

LESSON 14

Addition: Doubles +1

Associative Property

Once the doubles are mastered, we build on this knowledge with the doubles plus one. If you know 3 + 3 = 6, then you can see that 3 + 4 = 7. In the same way, think 4 + 5 = 4 + 4 + 1 = 9. Some call these particular facts the neighbors of the doubles. They say that 4 + 5 is the neighbor of 4 + 4 and 5 + 5. If you don't work it out as in the examples, think that 4 + 5 is between 4 + 4, which is 8, and 5 + 5, which is 10. Thus, 4 + 5 must be 9. Notice that we are not working by sheer rote memory; we are encouraging thinking and understanding. Both are necessary in learning and remembering math facts.

This is a good place to introduce the ***Associative Property of Addition***. Think of associating with two other friends. Pretend that only two people can associate at one time. The Associative Property of Addition states that you can regroup (or re-associate) the addends in an addition problem without affecting the sum.

In the following example, 3 + 4 = 7. First we decomposed the 4 as 3 + 1. Notice how 3 and 1 associate at first to make 4. Then the problem is regrouped to form 3 and 3, or 6, plus 1 more. The parentheses tell you how the numbers are being grouped, or associated.

Example 1

Solve: 3 + 4

3 + 4 = 3 + 3 + 1 = 6 + 1 = 7

3 + 4 = 3 + (3 + 1) = (3 + 3) + 1 = 6 + 1 = 7

Example 2
Solve: 5 + 6

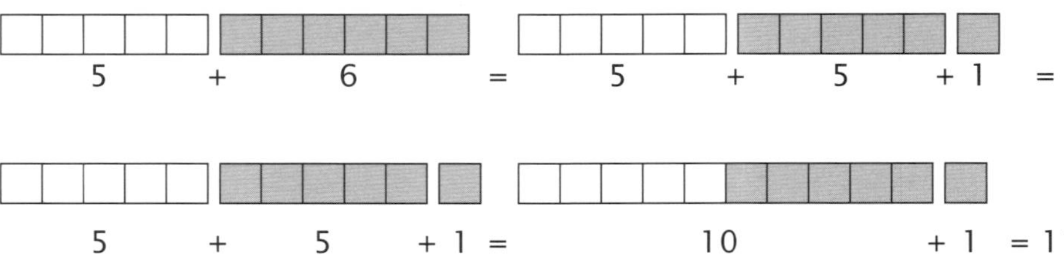

5 + 6 = 5 + (5 + 1) = (5 + 5) + 1 = 10 + 1 = 11

The doubles plus one are found above and below the doubles. There are eight new ones. Now we have accomplished all of the shaded facts on the chart. Eighty-eight out of 100 facts have been learned!

0+0	0+1	0+2	0+3	0+4	0+5	0+6	0+7	0+8	0+9
1+0	1+1	1+2	1+3	1+4	1+5	1+6	1+7	1+8	1+9
2+0	2+1	2+2	2+3	2+4	2+5	2+6	2+7	2+8	2+9
3+0	3+1	3+2	3+3	3+4	3+5	3+6	3+7	3+8	3+9
4+0	4+1	4+2	4+3	4+4	4+5	4+6	4+7	4+8	4+9
5+0	5+1	5+2	5+3	5+4	5+5	5+6	5+7	5+8	5+9
6+0	6+1	6+2	6+3	6+4	6+5	6+6	6+7	6+8	6+9
7+0	7+1	7+2	7+3	7+4	7+5	7+6	7+7	7+8	7+9
8+0	8+1	8+2	8+3	8+4	8+5	8+6	8+7	8+8	8+9
9+0	9+1	9+2	9+3	9+4	9+5	9+6	9+7	9+8	9+9

Sharing Parts—Introduction to Fractions

Most children learn the concept of simple fractions from life experience. It is a good idea to use the language ***one half*** and ***one fourth*** (as well as ***one quarter***) in everyday life. Also use the phrases "half of," "fourth of," and "quarter of" in conversation. Point out that when two people share equally, each gets one half, and each gets the same amount; however, the size of each portion is smaller than the original. If four people share equally, each one gets the same amount as the others. Each fourth is the same size, but it is smaller than a half of the same object.

Addition: Making 10

There are two ways to teach addition facts. I call the first the logical way, which is used to teach most of the facts. The second way is the family method, which we use for the 10 family. Take a 10 bar and place it on the table or floor. Using two of the colored unit bars, ask, "How many ways can you make 10?" This is referred to as decomposing ten in different ways. Be sure you verbally say "two unit bars" because addition facts are a combination of two numbers, not three or four different numbers. In Figure 1, you can see five ways of making 10. They are 1 + 9, 2 + 8, 3 + 7, 4 + 6, and 5 + 5. Our 10 family is made up of five different facts. After you build these, write them down and say them..

Figure 1

The ten facts and the nine facts (4 + 5, 6 + 3, 7 + 2, 8 + 1) are the most important ones to teach using the family method. Make sure these facts are "down cold." (Making nine is taught in the next lesson.)

Games for Making 10

Build a Wall - See who can build the highest wall that is 10 units long, using two bars in each row snapped onto the row beneath. After you build your wall, write down the equation that corresponds to the two pieces in each row. If the first row is a six bar and a four bar, then the equation is $6 + 4 = 10$.

Fill in the Space - Fill in the blank space with the correct bar. In this game you are still building on the 10, but instead of placing two bars to be the same length as 10, you place one bar, and the student places the other one. Then it is the student's turn to place one bar and your turn to figure out the length of the missing piece.

Example: Begin with a 10 bar and snap a seven bar on top of the 10 bar. Have the student find the piece that makes 10, in this case the three bar. Finally, have the student write down $7 + 3 = 10$. Then let the student be the teacher, while you find the missing unit bar and write the addition fact.

Race to a Hundred (and Race You Back) - Use either playing cards (with face cards removed and aces used as ones) or the cards you made for Pick a Card. Draw a card—for example, an eight. Pick up the eight bar and place it onto your 100 square. If your next card makes a sum less than 10, just add the block by the original bar. If the card makes a sum greater than 10, put the second bar somewhere else on the hundred square. On your next turn, pick another number and look for blocks that can even out a line, such as a two with an eight. Start at zero and accumulate blocks until you reach 100 and cover the entire 100 square.

When your student is comfortable with this, require each row of 10 to be filled in from top to bottom as they go. This forces dynamic addition/subtraction to be accomplished. By dynamic, I mean that blocks will have to be "traded in" to get the correct blocks that will fill in the required amount. Here's an example: the first number is seven, and second number is six. The six would be decomposed into two threes; one would fill in the first row along with the seven, and the second three would start the next row.

"Race You Back" means that students don't have to put their blocks away. In this game, you go in reverse and take off the value of the card. For the race back to zero, pull the number from anywhere within the hundred. A little dynamic work will be needed at the "100 point" and back near zero. Go as slowly as necessary. For example, there is only a five to go to be finished, and the student picks an eight. The eight must be decomposed into a five and a three.

Example 1

Solve: 3 + 7

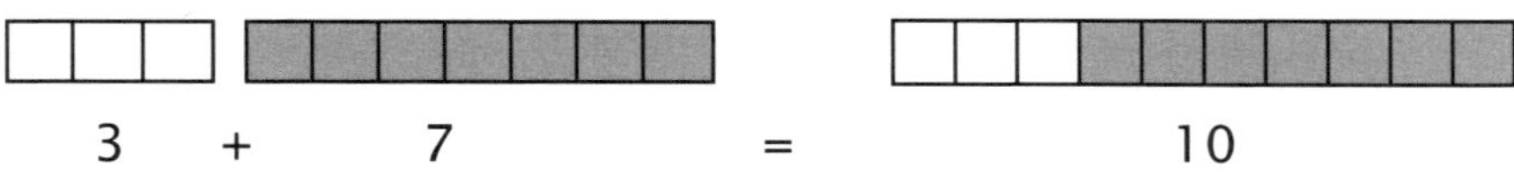

3 + 7 = 10

$3 + 7 = 10$
$7 + 3 = 10$

$$\begin{array}{r} 3 \\ +\ 7 \\ \hline 10 \end{array} \qquad \begin{array}{r} 7 \\ +\ 3 \\ \hline 10 \end{array}$$

Example 2

Solve: 6 + 4

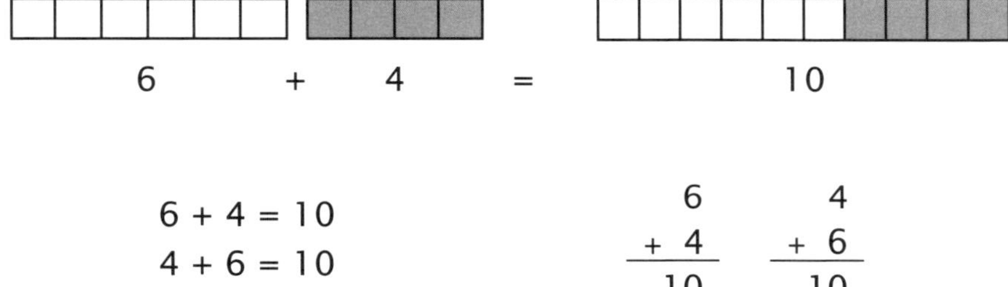

6 + 4 = 10

$6 + 4 = 10$
$4 + 6 = 10$

$$\begin{array}{r} 6 \\ +\ 4 \\ \hline 10 \end{array} \qquad \begin{array}{r} 4 \\ +\ 6 \\ \hline 10 \end{array}$$

We have emphasized that addition is to be taught with the different bars so that the student will learn the facts without counting. This lesson would be a good place to reinforce the concept that addition is fast counting. It may seem like a contradiction, but now that the students have learned so many of their facts, show them that they can find the answer to 5 + 5 by counting 1-2-3-4-5 and then 6-7-8-9-10. Show them that adding is faster and more efficient. Affirm that learning their facts without counting is more efficient. Provide encouragement for how well they are doing, reminding them that adding is fast counting.

Teaching Tip

Often a student will have difficulty with only a few facts. Choose one of them at a time and make it the "fact of the day." Post this fact on the refrigerator or some other place where it will be seen often. Then ask the student frequently during the day to tell you the answer. After talking about one fact all day long, you may find that the student knows that one best of all.

The making-10 facts cross the chart diagonally. There are four new ones. We have only eight more facts to go. Good work! Ninety-two out of 100 facts are behind us.

0+0	0+1	0+2	0+3	0+4	0+5	0+6	0+7	0+8	0+9
1+0	1+1	1+2	1+3	1+4	1+5	1+6	1+7	1+8	1+9
2+0	2+1	2+2	2+3	2+4	2+5	2+6	2+7	2+8	2+9
3+0	3+1	3+2	3+3	3+4	3+5	3+6	3+7	3+8	3+9
4+0	4+1	4+2	4+3	4+4	4+5	4+6	4+7	4+8	4+9
5+0	5+1	5+2	5+3	5+4	5+5	5+6	5+7	5+8	5+9
6+0	6+1	6+2	6+3	6+4	6+5	6+6	6+7	6+8	6+9
7+0	7+1	7+2	7+3	7+4	7+5	7+6	7+7	7+8	7+9
8+0	8+1	8+2	8+3	8+4	8+5	8+6	8+7	8+8	8+9
9+0	9+1	9+2	9+3	9+4	9+5	9+6	9+7	9+8	9+9

Mental Math

Here are some more questions to read to your student. Don't try the longer problems unless the student is comfortable with the shorter ones.

1. Three plus four, plus three equals what number? (10)
2. Six plus one, plus seven equals what number? (14)
3. One plus eight, plus eight equals what number? (17)
4. Two plus three, plus four equals what number? (9)
5. Nine plus zero, plus three equals what number? (12)
6. One plus one, plus two, plus three equals what number? (7)
7. Two plus one, plus zero, plus seven equals what number? (10)
8. Two plus two, plus one, plus six equals what number? (11)
9. Three plus one, plus four, plus one equals what number? (9)
10. Four plus four, plus one, plus five equals what number? (14)

LESSON 16

Addition: Making 9

Use the same technique for making nine as you did for making ten. Place a nine bar on the bottom and see how many ways you can make nine using two unit bars in each row. You will discover 1 + 8, 2 + 7, 3 + 6, and 5 + 4. The only new one is 3 + 6. We have already learned 1 + 8 in the one facts, 2 + 7 as one of the two facts, and 5 + 4 as a double plus one.

Figure 1

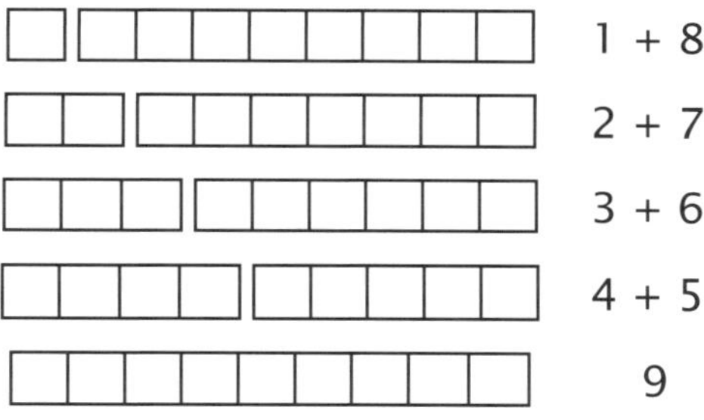

Knowing how to make nine will prove beneficial when learning to multiply. Make sure these facts are thoroughly mastered before proceeding to the next lesson.

Games for Making 9

Build a Wall - You may have played this game when learning the tens family. Let's adapt it to the nines. See who can build the highest wall nine units long, using two bars in each row snapped onto the previous row. As you build your wall, write down the equation that corresponds to the two pieces in each floor. If the first floor is a five bar and a four bar, write the equation 5 + 4 = 9 and say, "Five plus four equals nine."

Fill in the Space - Another game is to fill in the blank space with the correct bar. You are still building on the nine, but instead of placing two bars to be the same length as nine, you place one bar, and the student places the other one. Next it is the student's turn to place one bar while you figure out the length of the missing piece.

For example, begin with a nine bar, choose a seven bar, and snap it on top of the nine bar. Have the student find the piece that makes nine (in this case, the two bar,) and then write 7 + 2 = 9, saying, "Seven plus two equals nine." Change roles so that you find the missing unit bar and write the addition fact.

Wannabe 9 - Get out the one bar through the nine bar. Set the bars next to each other and ask how much more each needs in order to be a nine, or to "grow up to be a big nine."

Example 1

Solve: 3 + 6

3 + 6 = 9

3 + 6 = 9

$$\begin{array}{r} 3 \\ +\ 6 \\ \hline 9 \end{array}$$

Example 2

Solve: 6 + 3

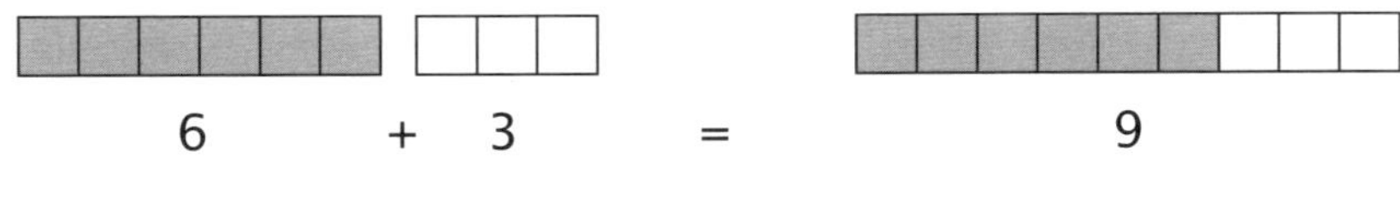

6 + 3 = 9

6 + 3 = 9

$$\begin{array}{r} 6 \\ +\ 3 \\ \hline 9 \end{array}$$

The making-nine facts are found diagonally above making ten. There are only two new ones. With only six facts to go, we can see the finish line. Don't give up now—you are almost done! Ninety-four out of 100 facts have been learned.

0+0	0+1	0+2	0+3	0+4	0+5	0+6	0+7	0+8	0+9
1+0	1+1	1+2	1+3	1+4	1+5	1+6	1+7	1+8	1+9
2+0	2+1	2+2	2+3	2+4	2+5	2+6	2+7	2+8	2+9
3+0	3+1	3+2	3+3	3+4	3+5	3+6	3+7	3+8	3+9
4+0	4+1	4+2	4+3	4+4	4+5	4+6	4+7	4+8	4+9
5+0	5+1	5+2	5+3	5+4	5+5	5+6	5+7	5+8	5+9
6+0	6+1	6+2	6+3	6+4	6+5	6+6	6+7	6+8	6+9
7+0	7+1	7+2	7+3	7+4	7+5	7+6	7+7	7+8	7+9
8+0	8+1	8+2	8+3	8+4	8+5	8+6	8+7	8+8	8+9
9+0	9+1	9+2	9+3	9+4	9+5	9+6	9+7	9+8	9+9

Teaching Tip

Go to mathusee.com for an interactive math-facts drill. It can be set to practice any combination of previously-learned facts. When you need a fresh approach to learning the facts, try inventing your own games using cards or dominoes.

Addition of the Extras: 3 + 5, 4 + 7, 5 + 7

The extras don't fit into a pattern like the eights or nines do, nor are they a part of the 9 or 10 families of math facts. They are also the last three facts to be learned. Actually, there are six facts, counting their commutative counterparts. Take your time and finish up the task of learning all of the addition facts.

Adding 3 + 5, 4 + 7, and 5 + 7 provides a good opportunity to use what students have already learned to figure out the new material. Looking at 4 + 7, they might say, "Well, that is one more than 3 + 7, which is 10, so the answer must be 11." Looking at 3 + 5 they might say, "Well, that is one less than 3 + 6, which is 9, so the answer must be 8." Give them a chance to figure out alternate strategies and encourage them to discover their own formulas and patterns. Help them understand that there are several ways to arrive at each answer.

Example 1

Solve: 3 + 5

3 + 5 = 8

$$3 + 5 = 8$$
$$5 + 3 = 8$$

$$\begin{array}{r} 3 \\ +\ 5 \\ \hline 8 \end{array} \qquad \begin{array}{r} 5 \\ +\ 3 \\ \hline 8 \end{array}$$

Example 2
Solve: 4 + 7

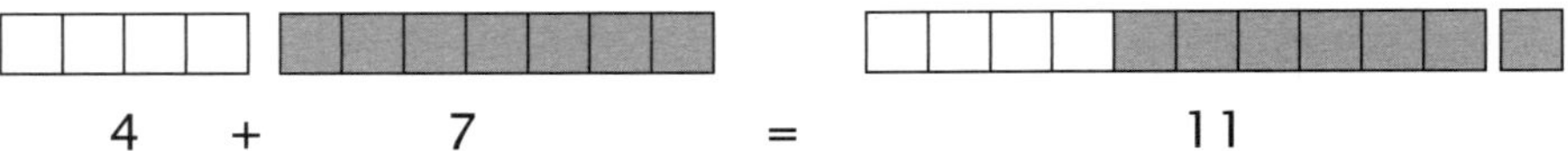

4 + 7 = 11

4 + 7 = 11
7 + 4 = 11

$$\begin{array}{r} 4 \\ +\ 7 \\ \hline 11 \end{array} \qquad \begin{array}{r} 7 \\ +\ 4 \\ \hline 11 \end{array}$$

Example 3
Solve: 5 + 7

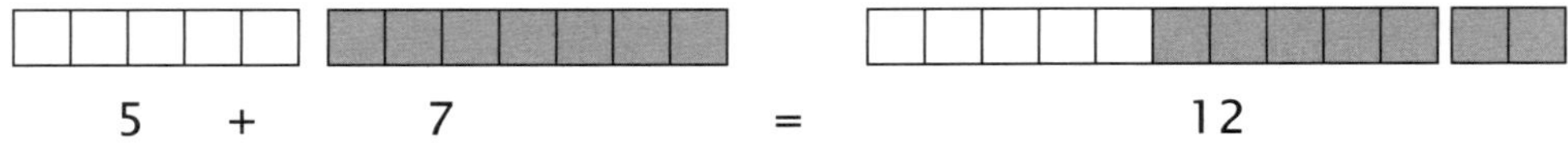

5 + 7 = 12

5 + 7 = 12
7 + 5 = 12

$$\begin{array}{r} 5 \\ +\ 7 \\ \hline 12 \end{array} \qquad \begin{array}{r} 7 \\ +\ 5 \\ \hline 12 \end{array}$$

This is the end of the individual addition facts. Please make sure all the facts are mastered before moving ahead to the next lesson. Having these facts under their belts will give students confidence and let them focus on the new material without continually backtracking and doing too many things at the same time. Take whatever time students need for mastery to occur.

An effective way to practice the facts is to use decomposition, as explained in lesson 10. Give the child a number and ask for two or more numbers that will add to make that number. Show the answer with the blocks and write it using an equation. Students should be able to decompose any number up to 10 in several different ways. For example, 9 could be expressed as 8 + 1, 4 + 5, or 5 + 3 + 1.

You did it! Congratulations on this accomplishment. After doing such a good job with addition, you have a strong foundation for learning subtraction. Well done!

0+0	0+1	0+2	0+3	0+4	0+5	0+6	0+7	0+8	0+9
1+0	1+1	1+2	1+3	1+4	1+5	1+6	1+7	1+8	1+9
2+0	2+1	2+2	2+3	2+4	2+5	2+6	2+7	2+8	2+9
3+0	3+1	3+2	3+3	3+4	3+5	3+6	3+7	3+8	3+9
4+0	4+1	4+2	4+3	4+4	4+5	4+6	4+7	4+8	4+9
5+0	5+1	5+2	5+3	5+4	5+5	5+6	5+7	5+8	5+9
6+0	6+1	6+2	6+3	6+4	6+5	6+6	6+7	6+8	6+9
7+0	7+1	7+2	7+3	7+4	7+5	7+6	7+7	7+8	7+9
8+0	8+1	8+2	8+3	8+4	8+5	8+6	8+7	8+8	8+9
9+0	9+1	9+2	9+3	9+4	9+5	9+6	9+7	9+8	9+9

Solving for the Unknown with Three Addends

In this book, some addition facts were taught by showing students how to break a number into smaller parts. For example, 6 + 4, 2 + 8, and 3 + 7 all make 10. Children who have lots of experience using the blocks should also be able to decompose numbers into three parts. (See lesson 10 teaching tip.) These skills can be used to solve slightly more challenging word problems. Notice that the unknown may appear in different locations in different problems.

1. Sue, Nancy, and Robin each picked flowers. Nancy picked three flowers, and Robin picked two flowers. Altogether, the girls had 10 flowers. How many flowers did Sue pick? $F + 3 + 2 = 10$

Students can see how to solve this by using the blocks. Put a three bar and a two bar on top of a 10 bar and see what is needed to fill the leftover space. Some students may think $3 + 2 = 5$, so they have the equation $F + 5 = 10$, which is a familiar problem. Use the blocks as much as needed to solve these problems.

2. Claire read one story to the baby on Thursday and two stories on Friday. After reading on Saturday, Claire had read eight stories altogether. How many stories did she read on Saturday? $1 + 2 + S = 8$

   Solving for the unknown using blocks or mental addition, we find that five stories were read on Saturday.

Teaching Tip

If you have a student who learns best when moving, try this tip. Take a child's ball and write the numbers 0–9 on it with an indelible marker. Toss the ball to the student. Then have the student add the numbers that her thumbs are touching. Visit mathusee.com for review tools and other resources.

LESSON 18

Introduction to Subtraction

After we've mastered place value and all the addition facts, we are ready for subtraction. ***Subtraction*** is the opposite, or inverse, of addition. If you know how to add (2 + 3 = 5) and how to solve for an unknown (X + 2 = 5 and X + 3 = 5), then subtraction will follow easily. Rather than teach another complete set of facts, we will relate subtraction to these two previously-mastered skills. If these skills are not mastered, please go back and spend whatever time is necessary until they are. Every subtraction problem can be rewritten and rephrased as an addition problem. We can read the example 5 – 2 = ? as "What number plus 2 is the same as 5?" This is identical to X + 2 = 5.

The answer to a subtraction problem is called the ***difference***. Instead of focusing on "taking away" or "minusing," focus on the difference between the two numbers being subtracted. This is why we can take away, and we can also add up.

Start by pushing the blocks end to end as you would for an addition problem. (See Example 1 on the next page.) Since subtraction is the opposite of addition, instead of leaving the blocks in that position, invert the two bar and place it on top of the five bar. The dark blocks in the picture represent the hollow side of the block. When the hollow side is showing, it means "take away," "minus," or "owe." One parent commented that you are "in the hole" when the hollow side is showing. Make a real-life problem by saying, "I have five dollars, and I owe my friend two dollars. How much do I have left?"

The symbol for subtraction is a single line. When inverting the bottom number and placing it on the top number, say "difference between." You could picture the single line representing subtraction as an arrow pointing out the "difference between."

Example 1

Solve: 5 - 2 or $\begin{array}{r} 5 \\ -\ 2 \\ \hline \end{array}$

Step 1

Rephrase as an addition problem.

"Five minus two" is the same as
"What plus two is the same as five?"

Step 2

Rewrite as solving for an unknown. X + 2 = 5

Step 3

Build as an addition problem.

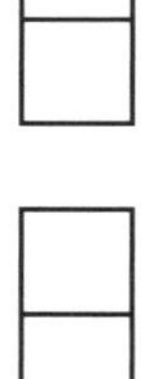

Step 4

Invert and reposition.

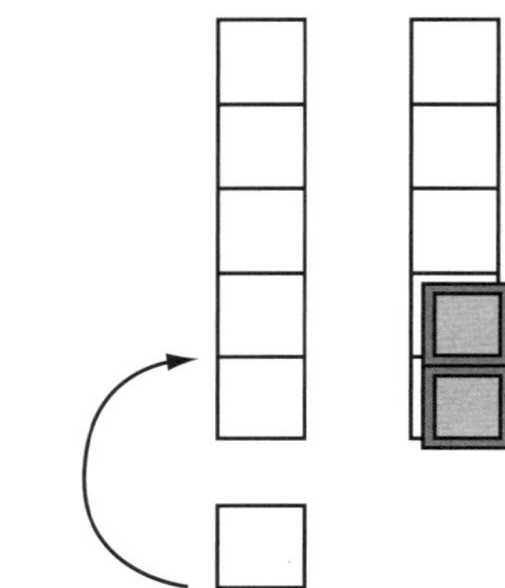

Step 5

Solve and write.

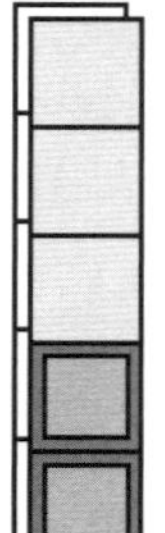

We see that 3 + 2 = 5;
thus, 5 - 2 = 3.

$$\begin{array}{r} 5 \\ -\ 2 \\ \hline 3 \end{array}$$

Example 2

Solve: 7 - 3 or $\begin{array}{r} 7 \\ -3 \\ \hline \end{array}$

Step 1

Rephrase as an addition problem.

"Seven minus three" means
"What plus three is the same as seven?"

Step 2

Rewrite as solving for an unknown. X + 3 = 7

Step 3

Build as an addition problem.

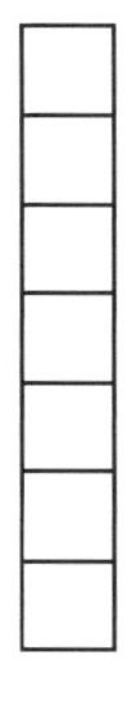

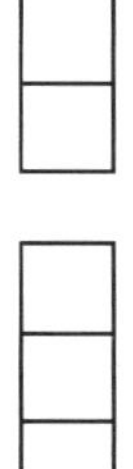

Step 4

Invert and reposition.

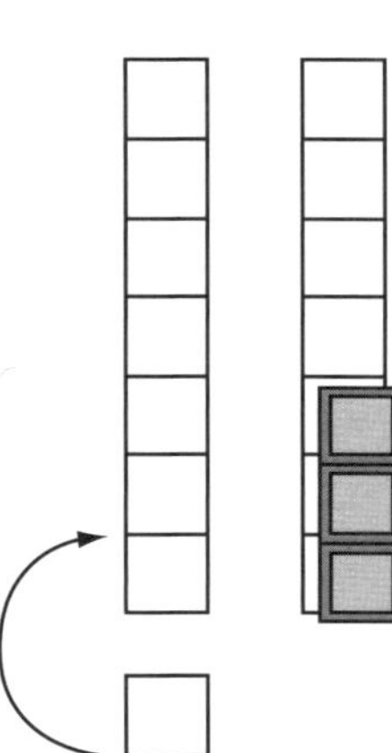

Step 5

Solve and write.

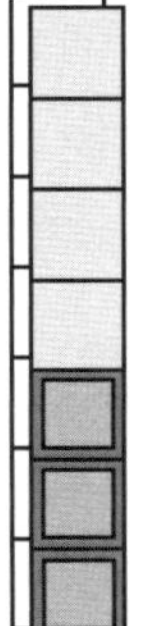

We see that 4 + 3 = 7;
thus, 7 - 3 = 4.

$$\begin{array}{r} 7 \\ -\ 3 \\ \hline 4 \end{array}$$

In this lesson, work on rewriting a subtraction problem as an addition problem. Use the blocks to show how subtraction is the inverse of addition. Help the student see how each problem can be solved by adding up. Now we see why it is so important to have the addition facts mastered at this point.

Remind students that they have already been subtracting as they solved for an unknown. This skill lays a foundation for algebra and reviews addition facts, but is also prepares students for subtraction. This should be an encouragement for those who have worked hard at learning this skill.

Mental Math

Here are some more questions to read to your student. All of these review addition. You may shorten these if your student is not yet ready for the longer mental math questions.

1. One plus two, plus three, plus four equals what number? (10)
2. Two plus two, plus one, plus two equals what number? (7)
3. Three plus one, plus four, plus one equals what number? (9)
4. Eight plus zero, plus one, plus three equals what number? (12)
5. Five plus one, plus one, plus five equals what number? (12)
6. Five plus four, plus zero, plus seven equals what number? (16)
7. Two plus two, plus two, plus two equals what number? (8)
8. Three plus four, plus one, plus five equals what number? (13)
9. Two plus three, plus two, plus three equals what number? (10)
10. Zero plus one, plus five, plus two equals what number? (8)

Subtraction: –1 and –0

As with addition, we'll begin with the simpler facts and move on from there. The first facts to be learned are the minus zero facts. If I have five dollars and I owe five dollars, how many dollars do I have? The answer is zero, or nothing. The problem is written as 5 – 5 = 0. Look at example 1 to see it worked through. Before teaching subtraction by one, review counting backwards by one with the blocks arranged from one to nine as we did when adding nine (lesson 9).

As you do the subtraction word problems, notice the different vocabulary, such as difference, more than, less than, fewer than, and have left. Remember to read for meaning, as not all subtraction problems will have these particular words.

Example 1

Solve: 5 - 5 or $\begin{array}{r} 5 \\ -5 \\ \hline \end{array}$

Step 1

Rephrase as an addition problem.

"Five minus five" means
"What plus five is the same as five?"

Step 2

Rewrite as solving for an unknown.

X + 5 = 5

Step 3
Build as an addition problem.

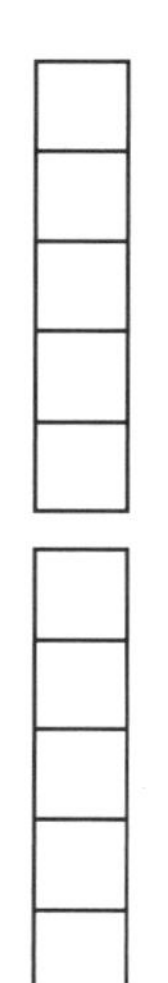

Step 4
Invert and reposition.

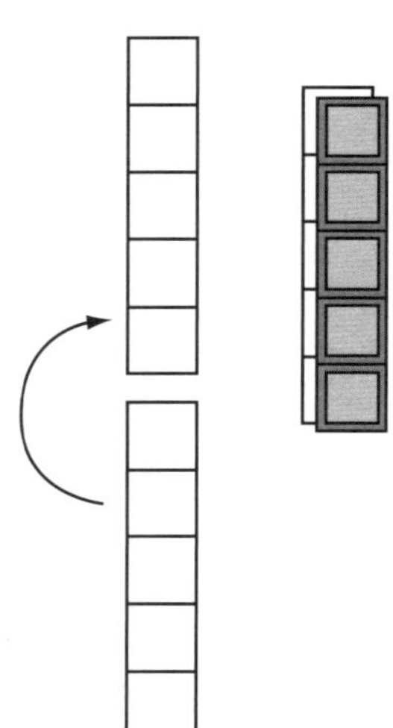

Step 5
Solve and write.

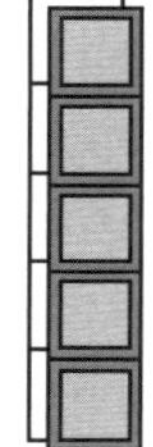

We see that 0 + 5 = 5;
thus, 5 - 5 = 0.

$$\begin{array}{r} 5 \\ -\ 5 \\ \hline 0 \end{array}$$

Subtraction is not commutative. We see that 5 – 3 is not the same as 3 – 5. Addition *is* commutative, so 5 + 3 is the same as 3 + 5. However, when we rewrite a subtraction problem as an addition problem, we can tap into the Commutative Property of Addition. In the problem 5 – 3 = 2, we can switch the 2 and the 3, since adding is commutative and we are adding up. Notice that 2 + 3 = 5 is just the same as 3 + 2 = 5, so we know that 5 – 2 = 3 is the same as 5 – 3 = 2.

Subtracting One

The prerequisite for learning how to find a difference of one is learning how to add by one. Study Example 2. When we add up, $4 - 3 = 1$ is the same as $4 - 1 = 3$.

Example 2

Solve: 4 - 3 or $\begin{array}{r} 4 \\ -3 \\ \hline \end{array}$

Step 1

Rephrase as an addition problem.

"Four minus three" means
"What plus three is the same as four?"

Step 2

Rewrite as solving for an unknown. $X + 3 = 4$

Step 3

Build as an addition problem.

Step 4

Invert and reposition.

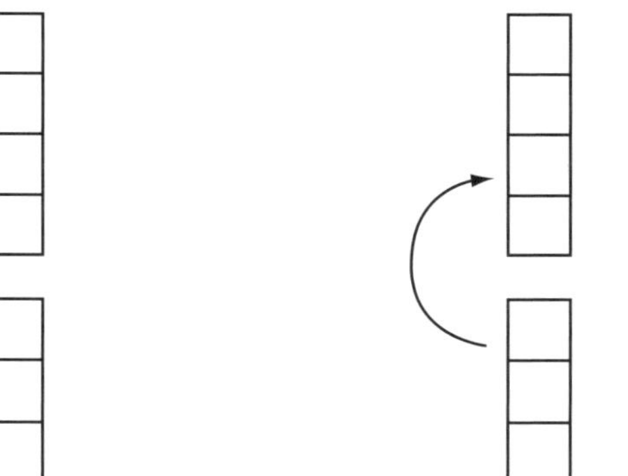

Step 5

Solve and write.

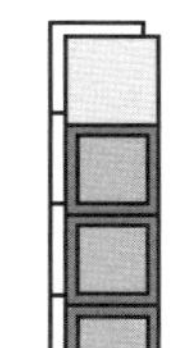

We see that $1 + 3 = 4$;
thus, $4 - 3 = 1$.

$$\begin{array}{r} 4 \\ -\ 1 \\ \hline 3 \end{array}$$

Step 5

The related fact for Example 2

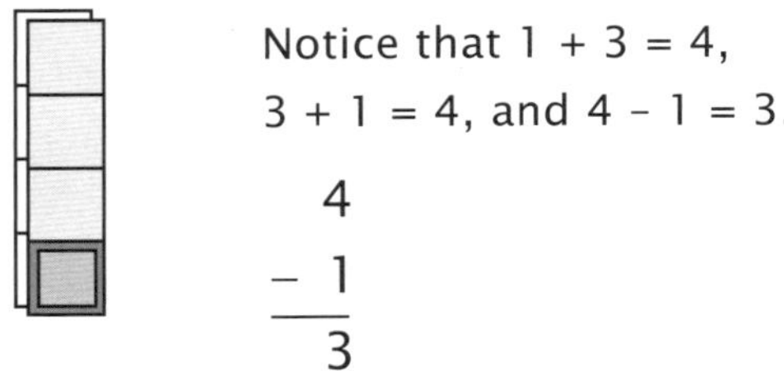

Notice that 1 + 3 = 4,
3 + 1 = 4, and 4 - 1 = 3.

$$\begin{array}{r} 4 \\ -\ 1 \\ \hline 3 \end{array}$$

Because of the Commutative Property of Addition, when we learn the zero facts, which are in the first column and written vertically, we also learn the first row, written horizontally. This is also true for the one facts, where we learn the second column as well as the second row. Make a habit of encouraging the student by showing what we are learning lesson by lesson. There is a chart for the student after lesson 19 in the student workbook.

0 - 0	1 - 1	2 - 2	3 - 3	4 - 4	5 - 5	6 - 6	7 - 7	8 - 8	9 - 9
1 - 0	2 - 1	3 - 2	4 - 3	5 - 4	6 - 5	7 - 6	8 - 7	9 - 8	10 - 9
2 - 0	3 - 1	4 - 2	5 - 3	6 - 4	7 - 5	8 - 6	9 - 7	10 - 8	11 - 9
3 - 0	4 - 1	5 - 2	6 - 3	7 - 4	8 - 5	9 - 6	10 - 7	11 - 8	12 - 9
4 - 0	5 - 1	6 - 2	7 - 3	8 - 4	9 - 5	10 - 6	11 - 7	12 - 8	13 - 9
5 - 0	6 - 1	7 - 2	8 - 3	9 - 4	10 - 5	11 - 6	12 - 7	13 - 8	14 - 9
6 - 0	7 - 1	8 - 2	9 - 3	10 - 4	11 - 5	12 - 6	13 - 7	14 - 8	15 - 9
7 - 0	8 - 1	9 - 2	10 - 3	11 - 4	12 - 5	13 - 6	14 - 7	15 - 8	16 - 9
8 - 0	9 - 1	10 - 2	11 - 3	12 - 4	13 - 5	14 - 6	15 - 7	16 - 8	17 - 9
9 - 0	10 - 1	11 - 2	12 - 3	13 - 4	14 - 5	15 - 6	16 - 7	17 - 8	18 - 9

Subtraction Facts Sheet

0 – 0	1 – 1	2 – 2	3 – 3	4 – 4	5 – 5	6 – 6	7 – 7	8 – 8	9 – 9
1 – 0	2 – 1	3 – 2	4 – 3	5 – 4	6 – 5	7 – 6	8 – 7	9 – 8	10 – 9
2 – 0	3 – 1	4 – 2	5 – 3	6 – 4	7 – 5	8 – 6	9 – 7	10 – 8	11 – 9
3 – 0	4 – 1	5 – 2	6 – 3	7 – 4	8 – 5	9 – 6	10 – 7	11 – 8	12 – 9
4 – 0	5 – 1	6 – 2	7 – 3	8 – 4	9 – 5	10 – 6	11 – 7	12 – 8	13 – 9
5 – 0	6 – 1	7 – 2	8 – 3	9 – 4	10 – 5	11 – 6	12 – 7	13 – 8	14 – 9
6 – 0	7 – 1	8 – 2	9 – 3	10 – 4	11 – 5	12 – 6	13 – 7	14 – 8	15 – 9
7 – 0	8 – 1	9 – 2	10 – 3	11 – 4	12 – 5	13 – 6	14 – 7	15 – 8	16 – 9
8 – 0	9 – 1	10 – 2	11 – 3	12 – 4	13 – 5	14 – 6	15 – 7	16 – 8	17 – 9
9 – 0	10 – 1	11 – 2	12 – 3	13 – 4	14 – 5	15 – 6	16 – 7	17 – 8	18 – 9

Subtraction: –2

In this lesson we are learning to subtract by two. Because of what we know about the Commutative Property of Addition, this lesson also includes those problems with a difference of two. Example 1 shows minus two, and Example 2 shows the same problem with a difference of two. In a subtraction problem the top number is called the ***minuend***, and the bottom number is the ***subtrahend***.

Figure 1

7	minuend
–2	subtrahend
5	difference

Example 1

Solve: 7 – 2 or $\begin{array}{r} 7 \\ -2 \\ \hline \end{array}$

Step 1

Rephrase as an addition problem.

"Seven minus two" means "What plus two is the same as seven?"

Step 2

Rewrite as solving for an unknown.

X + 2 = 7

Step 3
Build as an addition problem.

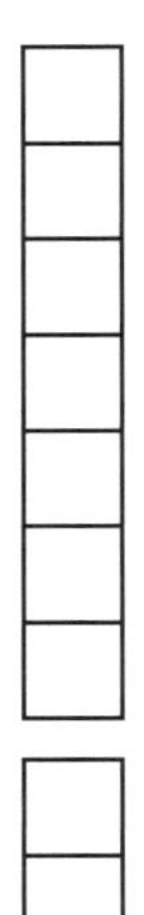

Step 4
Invert and reposition.

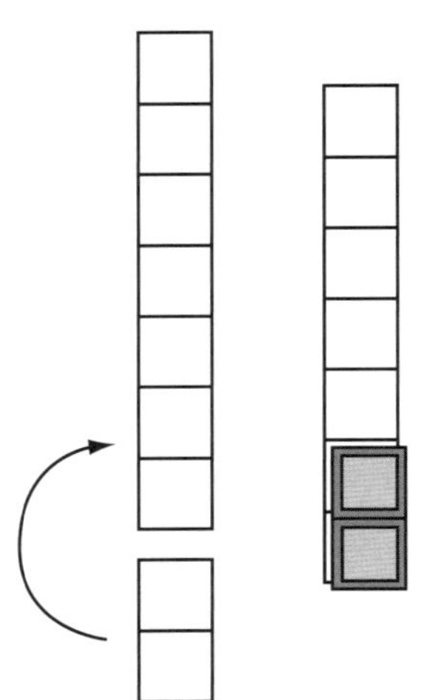

Step 5
Solve and write.

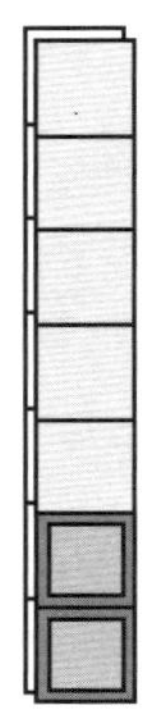

We see that 2 + 5 = 7;
thus, 7 - 2 = 5.

$$\begin{array}{r} 7 \\ -\ 2 \\ \hline 5 \end{array}$$

Example 2
Solve: 7 - 5 or $\begin{array}{r} 7 \\ -\ 5 \\ \hline \end{array}$

Step 1 and 2
Rephrase and rewrite.

"Seven minus five" means
"What plus five is the same as seven?" X + 5 = 7

Step 3
Build as an addition problem.

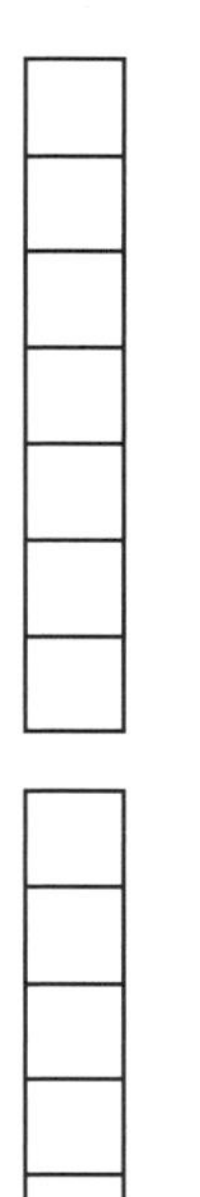

Step 4
Invert and reposition.

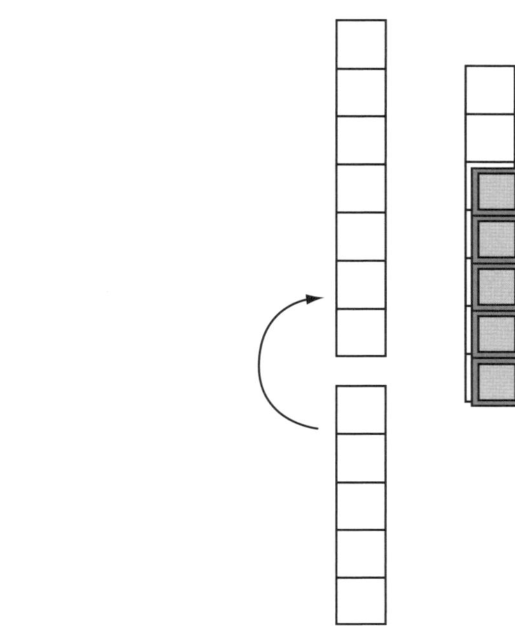

Step 5
Solve and write.

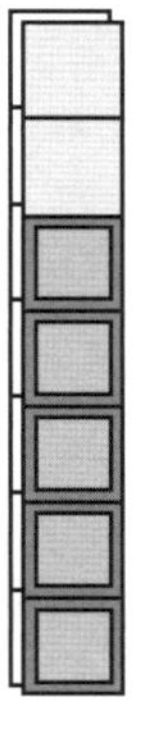

We see that 5 + 2 = 7;
thus, 7 - 5 = 2.

$$\begin{array}{r} 7 \\ -\ 5 \\ \hline 2 \end{array}$$

Because of the Commutative Property of Addition, when we learn the two facts, which are in the third column, we also learn the facts with a difference of two in the third row. We have already learned over half of the facts. Fifty-one have been mastered, and there are 49 left to learn. Congratulations!

0 - 0	1 - 1	2 - 2	3 - 3	4 - 4	5 - 5	6 - 6	7 - 7	8 - 8	9 - 9
1 - 0	2 - 1	3 - 2	4 - 3	5 - 4	6 - 5	7 - 6	8 - 7	9 - 8	10 - 9
2 - 0	3 - 1	4 - 2	5 - 3	6 - 4	7 - 5	8 - 6	9 - 7	10 - 8	11 - 9
3 - 0	4 - 1	5 - 2	6 - 3	7 - 4	8 - 5	9 - 6	10 - 7	11 - 8	12 - 9
4 - 0	5 - 1	6 - 2	7 - 3	8 - 4	9 - 5	10 - 6	11 - 7	12 - 8	13 - 9
5 - 0	6 - 1	7 - 2	8 - 3	9 - 4	10 - 5	11 - 6	12 - 7	13 - 8	14 - 9
6 - 0	7 - 1	8 - 2	9 - 3	10 - 4	11 - 5	12 - 6	13 - 7	14 - 8	15 - 9
7 - 0	8 - 1	9 - 2	10 - 3	11 - 4	12 - 5	13 - 6	14 - 7	15 - 8	16 - 9
8 - 0	9 - 1	10 - 2	11 - 3	12 - 4	13 - 5	14 - 6	15 - 7	16 - 8	17 - 9
9 - 0	10 - 1	11 - 2	12 - 3	13 - 4	14 - 5	15 - 6	16 - 7	17 - 8	18 - 9

Teaching Tip

Remember the importance of applying math to everyday life. Ask questions as you shop, cook, and do daily chores.

LESSON 21

Subtraction: –9

Subtracting by nine, which used to be one of the toughest sets of problems to learn, can now be one of the easiest. It is simply adding by one. This is a two-step problem. The first step is to make 10. Continue adding up to the number in the units place. The blocks clearly show this. We inserted an outline of a 10 to show both steps.

Example 1

$$\begin{array}{r} 13 \\ -\ 9 \\ \hline \end{array}$$

10 plus what equals 13?
The answer is 3.
9 plus what equals 10?
The answer is 1.

$$\begin{array}{r} 13 \\ 10 \\ -\ 9 \\ \hline 4 \end{array}$$

3
1

The difference (or distance between) 9 and 13 is 1 + 3, or 4.

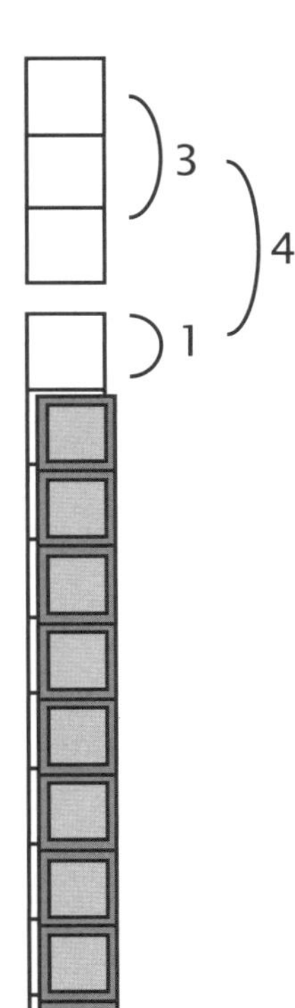

You may remember how we learned to add by nine. First we "vacuumed" one from the number and added 10. In subtraction, the opposite or inverse of addition, we add up by one instead of "vacuuming," or taking away one.

Example 2

$$\begin{array}{r} 15 \\ -\ 9 \\ \hline \end{array}$$

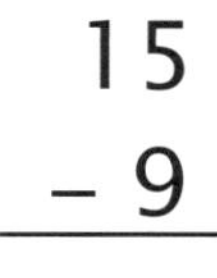

10 plus what equals 15?
The answer is 5.
9 plus what equals 10?
The answer is 1.

The difference (or distance between) 9 and 15 is 1 + 5, or 6.

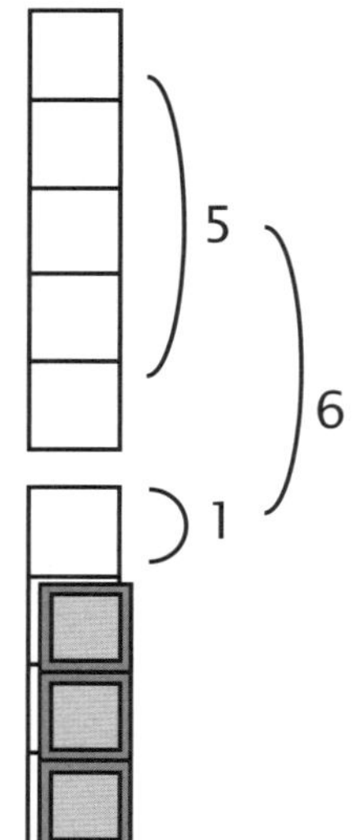

Example 3

$$\begin{array}{r} 16 \\ -\ 9 \\ \hline \end{array}$$

10 plus what equals 16?
The answer is 6.
9 plus what equals 10?
The answer is 1.

$$\begin{array}{rl} 16 & \\ 10 & \rceil\, 6 \\ -\ 9 & \rceil\, 1 \\ \hline 7 & \end{array}$$

The difference (or distance between) 9 and 16 is 1 + 6, or 7.

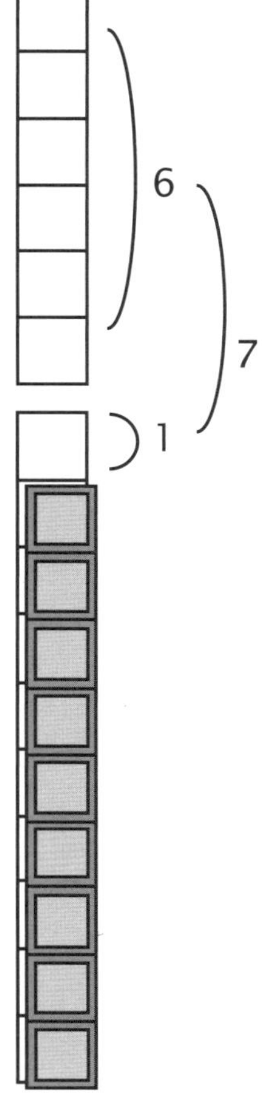

Now we have learned 58 facts with just 42 to go. Keep up the good work!

0 - 0	1 - 1	2 - 2	3 - 3	4 - 4	5 - 5	6 - 6	7 - 7	8 - 8	9 - 9
1 - 0	2 - 1	3 - 2	4 - 3	5 - 4	6 - 5	7 - 6	8 - 7	9 - 8	10 - 9
2 - 0	3 - 1	4 - 2	5 - 3	6 - 4	7 - 5	8 - 6	9 - 7	10 - 8	11 - 9
3 - 0	4 - 1	5 - 2	6 - 3	7 - 4	8 - 5	9 - 6	10 - 7	11 - 8	12 - 9
4 - 0	5 - 1	6 - 2	7 - 3	8 - 4	9 - 5	10 - 6	11 - 7	12 - 8	13 - 9
5 - 0	6 - 1	7 - 2	8 - 3	9 - 4	10 - 5	11 - 6	12 - 7	13 - 8	14 - 9
6 - 0	7 - 1	8 - 2	9 - 3	10 - 4	11 - 5	12 - 6	13 - 7	14 - 8	15 - 9
7 - 0	8 - 1	9 - 2	10 - 3	11 - 4	12 - 5	13 - 6	14 - 7	15 - 8	16 - 9
8 - 0	9 - 1	10 - 2	11 - 3	12 - 4	13 - 5	14 - 6	15 - 7	16 - 8	17 - 9
9 - 0	10 - 1	11 - 2	12 - 3	13 - 4	14 - 5	15 - 6	16 - 7	17 - 8	18 - 9

Mental Math

Here are some more questions to read to your student. All of these review addition. You may shorten these if your student is not yet ready to do the longer questions mentally.

1. Four plus two, plus two, plus seven equals what number? (15)
2. Five plus three, plus one, plus three equals what number? (12)
3. Seven plus one, plus two, plus one equals what number? (11)
4. Three plus four, plus two, plus zero equals what number? (9)
5. One plus five, plus two, plus six equals what number? (14)
6. Zero plus one, plus three, plus three equals what number? (7)
7. Three plus three, plus three, plus three equals what number? (12)
8. Two plus four, plus one, plus nine equals what number? (16)
9. Three plus five, plus two, plus four equals what number? (14)
10. Four plus zero, plus eight, plus one equals what number? (13)
11. Six plus two, plus one, plus one equals what number? (10)
12. Three plus three, plus one, plus eight equals what number? (15)

LESSON 22

Subtraction: –8

Subtracting by eight is just like subtracting by nine, except that we need to add two when adding up. It is a two-step problem as well. The first step is to make 10. Then continue adding up to the digit in the units place. When we learned to add by eight, we "vacuumed" two from the number and then added 10. In subtraction by eight, we add up by two, instead of "vacuuming" two.

Example 1

$$\begin{array}{r} 13 \\ -\ 9 \\ \hline \end{array}$$

10 plus what equals 13?
The answer is 3.
9 plus what equals 10?
The answer is 1.

$$\begin{array}{rl} 13 & \\ 10 &) \; 3 \\ -\ 9 &) \; 1 \\ \hline 4 & \end{array}$$

The difference (or distance between) 9 and 13 is 1 + 3, or 4.

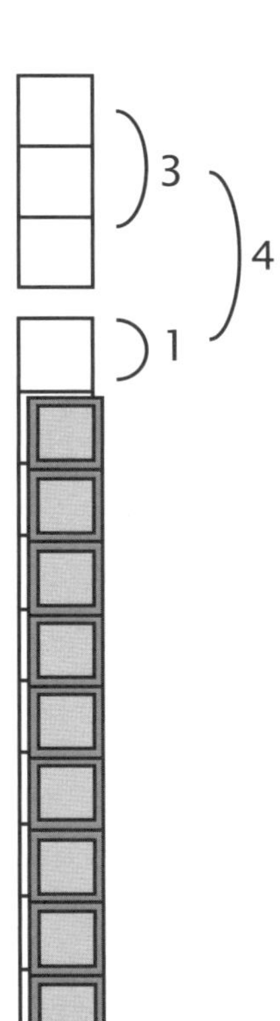

Example 2

$$\begin{array}{r} 11 \\ -\ 8 \\ \hline \end{array}$$

$$\begin{array}{r} 11 \\ 10 \\ -\ 8 \\ \hline 3 \end{array} \begin{array}{l} 1 \\ 2 \end{array}$$

10 plus what equals 11?
The answer is 1.
8 plus what equals 10?
The answer is 2.

The difference between 8 and 11 is 2 + 1, or 3.

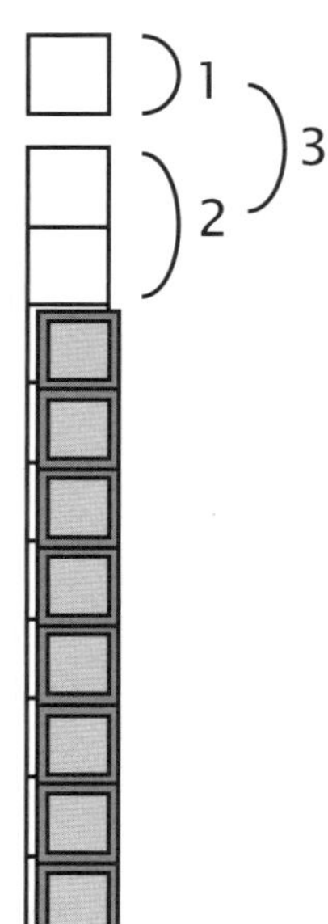

Example 3

$$\begin{array}{r} 15 \\ -\ 8 \\ \hline \end{array}$$

$$\begin{array}{r} 15 \\ 10 \\ -\ 8 \\ \hline 7 \end{array} \begin{array}{l} 5 \\ 2 \end{array}$$

10 plus what equals 15?
The answer is 5.
8 plus what equals 10?
The answer is 2.

The difference between 8 and 15 is 2 + 5, or 7.

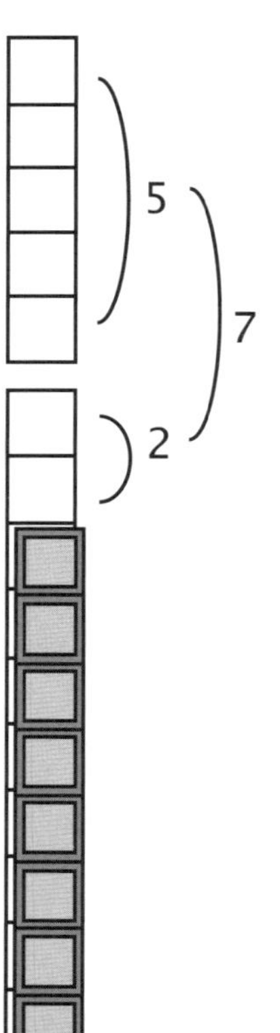

With the eights mastered, you have now learned 65 facts, with just 35 to go. Notice how the chart is almost all shaded. Be encouraged!

0 - 0	1 - 1	2 - 2	3 - 3	4 - 4	5 - 5	6 - 6	7 - 7	8 - 8	9 - 9
1 - 0	2 - 1	3 - 2	4 - 3	5 - 4	6 - 5	7 - 6	8 - 7	9 - 8	10 - 9
2 - 0	3 - 1	4 - 2	5 - 3	6 - 4	7 - 5	8 - 6	9 - 7	10 - 8	11 - 9
3 - 0	4 - 1	5 - 2	6 - 3	7 - 4	8 - 5	9 - 6	10 - 7	11 - 8	12 - 9
4 - 0	5 - 1	6 - 2	7 - 3	8 - 4	9 - 5	10 - 6	11 - 7	12 - 8	13 - 9
5 - 0	6 - 1	7 - 2	8 - 3	9 - 4	10 - 5	11 - 6	12 - 7	13 - 8	14 - 9
6 - 0	7 - 1	8 - 2	9 - 3	10 - 4	11 - 5	12 - 6	13 - 7	14 - 8	15 - 9
7 - 0	8 - 1	9 - 2	10 - 3	11 - 4	12 - 5	13 - 6	14 - 7	15 - 8	16 - 9
8 - 0	9 - 1	10 - 2	11 - 3	12 - 4	13 - 5	14 - 6	15 - 7	16 - 8	17 - 9
9 - 0	10 - 1	11 - 2	12 - 3	13 - 4	14 - 5	15 - 6	16 - 7	17 - 8	18 - 9

Teaching Tip

If a student has gotten into the habit of counting on fingers, it can be very hard to break. Some students know their facts but lack confidence. Here is a strategy that has helped some parents and teachers in this situation. Put the blocks on a different table, far from the one on which the student is working. Tell the student that he or she may not use fingers. The blocks may be used if the student is unsure of an answer. However, getting up to retrieve the blocks is a nuisance. For many students this is a way to encourage them to use what they really do know while giving them a way to find the answer if they need help.

LESSON 23

Subtraction: Doubles

Adding the doubles comes fairly quickly to many students. We hope subtracting doubles will be learned just as easily by your student. The difficult fact for adding is 7 + 7, but if you've learned that one, then you are well on your way to mastering the doubles facts.

Example 1

Solve: 10 - 5

$$\begin{array}{r} 10 \\ -\ 5 \\ \hline \end{array}$$

"Ten minus five" means "What plus five is the same as ten?"

X + 5 = 10

$$\begin{array}{r} 10 \\ -\ 5 \\ \hline 5 \end{array}$$

Five plus five equals ten.

5 + 5 = 10

Example 2

Solve: 8 - 4

$\begin{array}{r} 8 \\ -\ 4 \\ \hline \end{array}$ "Eight minus four" means "What plus four is the same as eight?"

$\begin{array}{r} 8 \\ -4 \\ \hline 4 \end{array}$ Four plus four equals eight. $4 + 4 = 8$

Example 3

Solve: 14 - 7

$\begin{array}{r} 14 \\ -\ 7 \\ \hline \end{array}$ "Fourteen minus seven" means "What plus seven is equal to fourteen?" $Y + 7 = 14$

$\begin{array}{r} 14 \\ -7 \\ \hline 7 \end{array}$ Seven plus seven equals fourteen. $7 + 7 = 14$

There are five new doubles facts, which makes 70 facts down and 30 facts to go. Notice that the doubles go diagonally on the chart. Keep up the good work!

0 - 0	1 - 1	2 - 2	3 - 3	4 - 4	5 - 5	6 - 6	7 - 7	8 - 8	9 - 9
1 - 0	2 - 1	3 - 2	4 - 3	5 - 4	6 - 5	7 - 6	8 - 7	9 - 8	10 - 9
2 - 0	3 - 1	4 - 2	5 - 3	6 - 4	7 - 5	8 - 6	9 - 7	10 - 8	11 - 9
3 - 0	4 - 1	5 - 2	6 - 3	7 - 4	8 - 5	9 - 6	10 - 7	11 - 8	12 - 9
4 - 0	5 - 1	6 - 2	7 - 3	8 - 4	9 - 5	10 - 6	11 - 7	12 - 8	13 - 9
5 - 0	6 - 1	7 - 2	8 - 3	9 - 4	10 - 5	11 - 6	12 - 7	13 - 8	14 - 9
6 - 0	7 - 1	8 - 2	9 - 3	10 - 4	11 - 5	12 - 6	13 - 7	14 - 8	15 - 9
7 - 0	8 - 1	9 - 2	10 - 3	11 - 4	12 - 5	13 - 6	14 - 7	15 - 8	16 - 9
8 - 0	9 - 1	10 - 2	11 - 3	12 - 4	13 - 5	14 - 6	15 - 7	16 - 8	17 - 9
9 - 0	10 - 1	11 - 2	12 - 3	13 - 4	14 - 5	15 - 6	16 - 7	17 - 8	18 - 9

LESSON 24

Subtraction: Making 10

The facts we cover in this lesson are directly related to the addition facts, just as all the other subtraction facts have been. Making the 10 family is the foundation for this set of subtraction facts. We know that 1 + 9, 2 + 8, 3 + 7, 4 + 6, and 5 + 5 are all the ways to make 10. We learned these facts by stacking the unit bars two in a row on top of the 10 bar. This is the inverse. Given one of the blocks, what will combine with it to make 10? Study the examples and have fun making 10. There are four new facts in this lesson.

Example 1

Solve: 10 - 3

$$\begin{array}{r} 10 \\ -3 \\ \hline \end{array}$$

"Ten minus three" means "What plus three is the same as ten?"

P + 3 = 10

$$\begin{array}{r} 10 \\ -3 \\ \hline 7 \end{array}$$

Seven plus three equals ten.

7 + 3 = 10

Example 2
Solve: 10 - 4

$$\begin{array}{r} 10 \\ -\ 4 \\ \hline \end{array}$$

"Ten minus four" means "What plus four is the same as ten?"

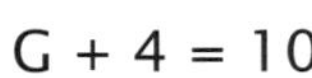

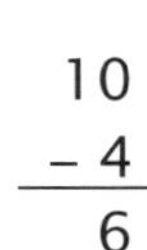

Six plus four equals ten.

6 + 4 = 10

Example 3
Solve: 10 - 7

$$\begin{array}{r} 10 \\ -\ 7 \\ \hline \end{array}$$

"Ten minus seven" means "What plus seven is the same as ten?"

P + 7 = 10

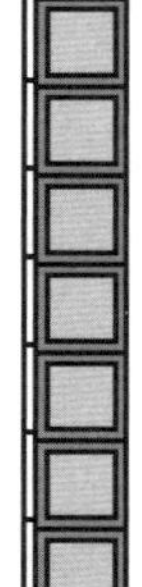

$$\begin{array}{r} 10 \\ -\ 7 \\ \hline 3 \end{array}$$

Three plus seven equals ten.

3 + 7 = 10

Example 3 is the related fact for Example 1. Do you see it?

There are four new facts. These appear diagonally, like the doubles do. We have conquered 74 facts, and 26 more are ready to be conquered.

0 - 0	1 - 1	2 - 2	3 - 3	4 - 4	5 - 5	6 - 6	7 - 7	8 - 8	9 - 9
1 - 0	2 - 1	3 - 2	4 - 3	5 - 4	6 - 5	7 - 6	8 - 7	9 - 8	10 - 9
2 - 0	3 - 1	4 - 2	5 - 3	6 - 4	7 - 5	8 - 6	9 - 7	10 - 8	11 - 9
3 - 0	4 - 1	5 - 2	6 - 3	7 - 4	8 - 5	9 - 6	10 - 7	11 - 8	12 - 9
4 - 0	5 - 1	6 - 2	7 - 3	8 - 4	9 - 5	10 - 6	11 - 7	12 - 8	13 - 9
5 - 0	6 - 1	7 - 2	8 - 3	9 - 4	10 - 5	11 - 6	12 - 7	13 - 8	14 - 9
6 - 0	7 - 1	8 - 2	9 - 3	10 - 4	11 - 5	12 - 6	13 - 7	14 - 8	15 - 9
7 - 0	8 - 1	9 - 2	10 - 3	11 - 4	12 - 5	13 - 6	14 - 7	15 - 8	16 - 9
8 - 0	9 - 1	10 - 2	11 - 3	12 - 4	13 - 5	14 - 6	15 - 7	16 - 8	17 - 9
9 - 0	10 - 1	11 - 2	12 - 3	13 - 4	14 - 5	15 - 6	16 - 7	17 - 8	18 - 9

Mental Math

Here are some more questions to read to your student. These combine addition and subtraction. You may need to go slowly with these at first.

1. Three plus four, minus two, equals what number? (5)
2. Six minus one, plus five, equals what number? (10)
3. One plus eight, minus eight, equals what number? (1)
4. Four minus two, plus nine, equals what number? (11)
5. Nine minus seven, plus four, equals what number? (6)
6. Eight minus five, plus seven, equals what number? (10)
7. Three plus eight, minus four, equals what number? (7)
8. Seven plus seven, minus five, equals what number? (9)
9. Eleven minus three, plus five, equals what number? (13)
10. Fifteen minus eight, plus one, equals what number? (8)

LESSON 25

Subtraction: Making 9

Making the nine family is the foundation for this set of subtraction facts. We know that 1 + 8, 2 + 7, 3 + 6, and 4 + 5 are all ways to make nine.

There are four new facts in this lesson. We will learn 9 – 3 and its related fact, 9 – 6, and 9 – 4 and its related fact, 9 – 5.

Example 1

Solve: 9 - 3

$$\begin{array}{r} 9 \\ -\ 3 \\ \hline \end{array}$$

"Nine minus three" means "What plus three is the same as nine?"

R + 3 = 9

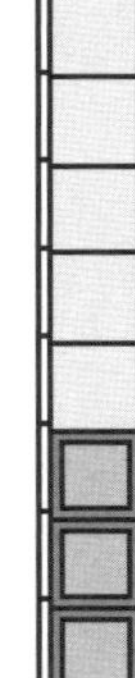

$$\begin{array}{r} 9 \\ -\ 3 \\ \hline 6 \end{array}$$

Six plus three equals nine.

6 + 3 = 9

Example 2
Solve: 9 - 4

$$\begin{array}{r} 9 \\ -\ 4 \\ \hline \end{array}$$

"Nine minus four" means "What plus four is the same as nine?"

R + 4 = 9

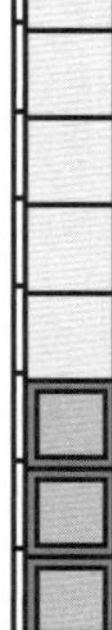

$$\begin{array}{r} 9 \\ -\ 4 \\ \hline 5 \end{array}$$

Five plus four equals nine.

5 + 4 = 9

Example 3 is the related fact for Example 2.

Example 3
Solve: 9 - 5

$$\begin{array}{r} 9 \\ -\ 5 \\ \hline \end{array}$$

"Nine minus five" means "What plus five is the same as nine?"

T + 5 = 9

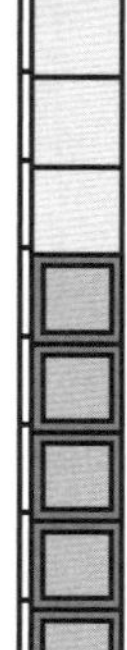

$$\begin{array}{r} 9 \\ -5 \\ \hline 4 \end{array}$$

Four plus five equals nine.

4 + 5 = 9

There are four new nine facts. These appear diagonally right above the 10 family. Seventy-eight facts are mastered, and 22 are ready to be learned.

0 - 0	1 - 1	2 - 2	3 - 3	4 - 4	5 - 5	6 - 6	7 - 7	8 - 8	9 - 9
1 - 0	2 - 1	3 - 2	4 - 3	5 - 4	6 - 5	7 - 6	8 - 7	9 - 8	10 - 9
2 - 0	3 - 1	4 - 2	5 - 3	6 - 4	7 - 5	8 - 6	9 - 7	10 - 8	11 - 9
3 - 0	4 - 1	5 - 2	6 - 3	7 - 4	8 - 5	9 - 6	10 - 7	11 - 8	12 - 9
4 - 0	5 - 1	6 - 2	7 - 3	8 - 4	9 - 5	10 - 6	11 - 7	12 - 8	13 - 9
5 - 0	6 - 1	7 - 2	8 - 3	9 - 4	10 - 5	11 - 6	12 - 7	13 - 8	14 - 9
6 - 0	7 - 1	8 - 2	9 - 3	10 - 4	11 - 5	12 - 6	13 - 7	14 - 8	15 - 9
7 - 0	8 - 1	9 - 2	10 - 3	11 - 4	12 - 5	13 - 6	14 - 7	15 - 8	16 - 9
8 - 0	9 - 1	10 - 2	11 - 3	12 - 4	13 - 5	14 - 6	15 - 7	16 - 8	17 - 9
9 - 0	10 - 1	11 - 2	12 - 3	13 - 4	14 - 5	15 - 6	16 - 7	17 - 8	18 - 9

LESSON 26

Subtraction: Extras

When we learned these facts in addition, they were referred to as the extras because they didn't fit into a pattern or a family. There are four new facts to be learned in this lesson, but because of the Commutative Property of Addition there are actually only two new ones to be learned. This lesson includes 7 – 4, 7 – 3, 8 – 5, and 8 – 3. These are the first subtraction facts taught without a family or some sort of easy pattern, so be sure you understand them before moving on.

Example 1

Solve: 7 - 3

$$\begin{array}{r} 7 \\ -\ 3 \\ \hline \end{array}$$

"Seven minus three" means "What plus three is the same as seven?"

$R + 3 = 7$

$$\begin{array}{r} 7 \\ -3 \\ \hline 4 \end{array}$$

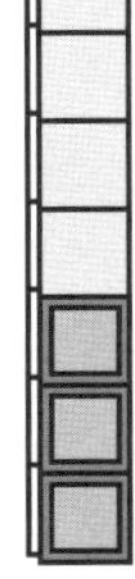

Four plus three equals seven.

$4 + 3 = 7$

Example 2

Solve: 7 - 4

$$\begin{array}{r} 7 \\ -\ 4 \\ \hline \end{array}$$

"Seven minus four" means "What plus four is the same as seven?"

C + 4 = 7

$$\begin{array}{r} 7 \\ -\ 4 \\ \hline 3 \end{array}$$

Three plus four equals seven.

3 + 4 = 7

Example 2 is related to Example 1. The related fact for Example 3 is 8 – 3 = 5.

Example 3

Solve: 8 - 5

$$\begin{array}{r} 8 \\ -\ 5 \\ \hline \end{array}$$

"Eight minus five" means "What plus five is the same as eight?"

Q + 5 = 8

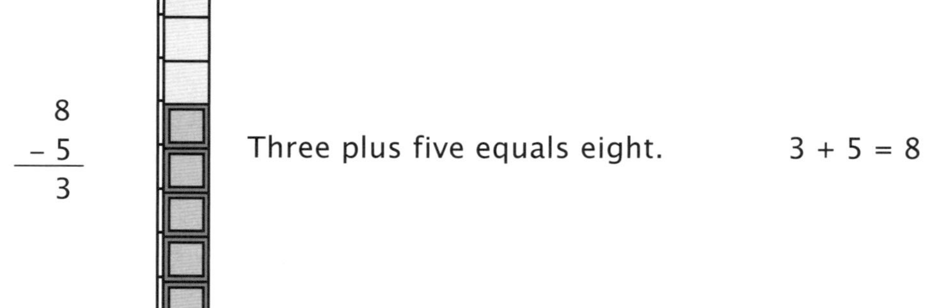

$$\begin{array}{r} 8 \\ -\ 5 \\ \hline 3 \end{array}$$

Three plus five equals eight.

3 + 5 = 8

We learned four new facts in this lesson. These appear in the corners above the 9 and 10 families. This lesson makes a grand total of 82 facts learned with only 18 facts left.

0 - 0	1 - 1	2 - 2	3 - 3	4 - 4	5 - 5	6 - 6	7 - 7	8 - 8	9 - 9
1 - 0	2 - 1	3 - 2	4 - 3	5 - 4	6 - 5	7 - 6	8 - 7	9 - 8	10 - 9
2 - 0	3 - 1	4 - 2	5 - 3	6 - 4	7 - 5	8 - 6	9 - 7	10 - 8	11 - 9
3 - 0	4 - 1	5 - 2	6 - 3	7 - 4	8 - 5	9 - 6	10 - 7	11 - 8	12 - 9
4 - 0	5 - 1	6 - 2	7 - 3	8 - 4	9 - 5	10 - 6	11 - 7	12 - 8	13 - 9
5 - 0	6 - 1	7 - 2	8 - 3	9 - 4	10 - 5	11 - 6	12 - 7	13 - 8	14 - 9
6 - 0	7 - 1	8 - 2	9 - 3	10 - 4	11 - 5	12 - 6	13 - 7	14 - 8	15 - 9
7 - 0	8 - 1	9 - 2	10 - 3	11 - 4	12 - 5	13 - 6	14 - 7	15 - 8	16 - 9
8 - 0	9 - 1	10 - 2	11 - 3	12 - 4	13 - 5	14 - 6	15 - 7	16 - 8	17 - 9
9 - 0	10 - 1	11 - 2	12 - 3	13 - 4	14 - 5	15 - 6	16 - 7	17 - 8	18 - 9

LESSON 27

Subtraction by 7, or Adding Up by 3

In the next four lessons, the minuend, or total, is a double-digit number. We are going to treat these as we did subtracting by eight and nine. This will be the two-step approach: first adding up to 10 and then adding the number in the units place to find the difference. We are assuming a thorough mastery of the 10 family for the remainder of the lessons dealing with subtraction facts. The facts to be learned in this lesson are 11 – 7, 12 – 7, 13 – 7, 15 – 7, and 16 – 7. Instead of subtracting by seven, we will be adding up by three.

Example 1

$$\begin{array}{r} 12 \\ -\ 7 \\ \hline \end{array}$$

10 plus what equals 12?
The answer is 2.
7 plus what equals 10?
The answer is 3.

$$\begin{array}{r} 12 \\ 10 \\ -\ 7 \\ \hline 5 \end{array}$$

The difference between 7 and 12 is 3 + 2, or 5.

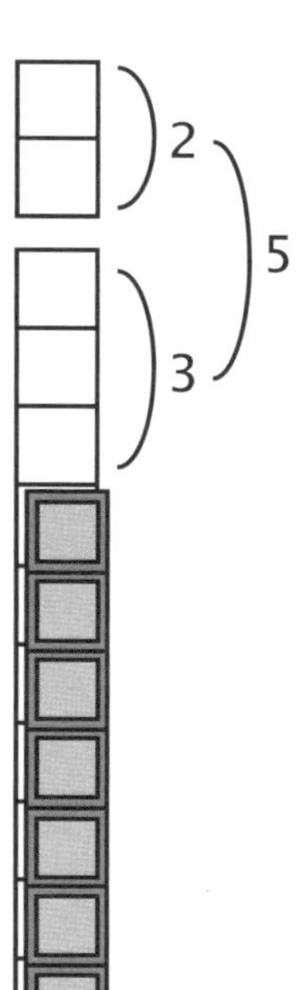

Example 2

$$\begin{array}{r} 13 \\ -\ 7 \\ \hline \end{array}$$

10 plus what equals 13?
The answer is 3.
7 plus what equals 10?
The answer is 3.

$$\begin{array}{r} 13 \\ 10 \\ -\ 7 \\ \hline 6 \end{array}$$

(13 to 10: 3; 10 to 7: 3)

The difference between 7 and 13 is 3 + 3, or 6.

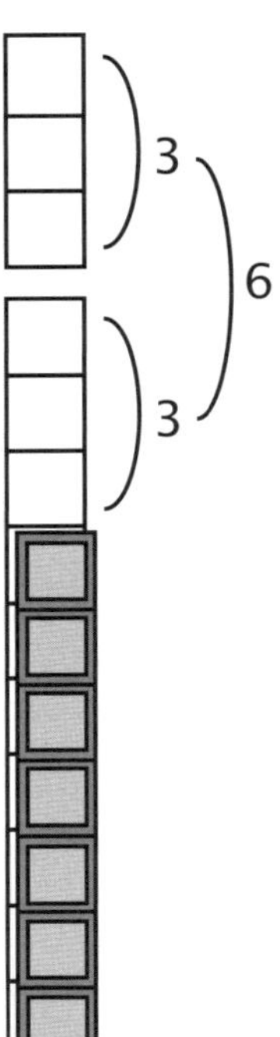

Example 3

$$\begin{array}{r} 15 \\ -\ 7 \\ \hline \end{array}$$

10 plus what equals 15?
The answer is 5.
7 plus what equals 10?
The answer is 3.

$$\begin{array}{r} 15 \\ 10 \\ -\ 7 \\ \hline 8 \end{array}$$

(15 to 10: 5; 10 to 7: 3)

The difference between 7 and 15 is 3 + 5, or 8.

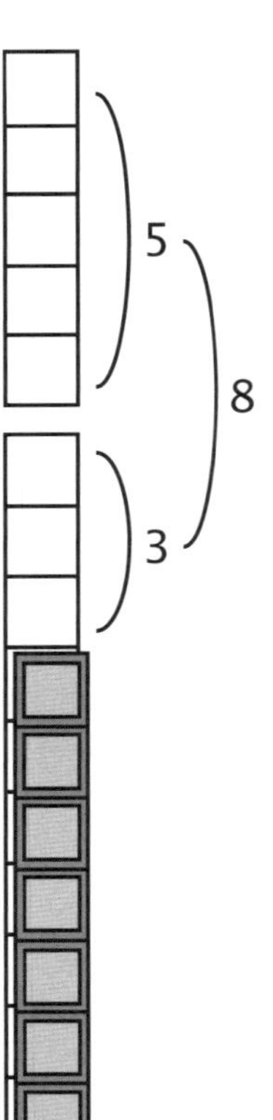

With the five new facts in this lesson, 87 have been mastered, and only 13 remain. The new group is in the minus seven column.

0 - 0	1 - 1	2 - 2	3 - 3	4 - 4	5 - 5	6 - 6	7 - 7	8 - 8	9 - 9
1 - 0	2 - 1	3 - 2	4 - 3	5 - 4	6 - 5	7 - 6	8 - 7	9 - 8	10 - 9
2 - 0	3 - 1	4 - 2	5 - 3	6 - 4	7 - 5	8 - 6	9 - 7	10 - 8	11 - 9
3 - 0	4 - 1	5 - 2	6 - 3	7 - 4	8 - 5	9 - 6	10 - 7	11 - 8	12 - 9
4 - 0	5 - 1	6 - 2	7 - 3	8 - 4	9 - 5	10 - 6	11 - 7	12 - 8	13 - 9
5 - 0	6 - 1	7 - 2	8 - 3	9 - 4	10 - 5	11 - 6	12 - 7	13 - 8	14 - 9
6 - 0	7 - 1	8 - 2	9 - 3	10 - 4	11 - 5	12 - 6	13 - 7	14 - 8	15 - 9
7 - 0	8 - 1	9 - 2	10 - 3	11 - 4	12 - 5	13 - 6	14 - 7	15 - 8	16 - 9
8 - 0	9 - 1	10 - 2	11 - 3	12 - 4	13 - 5	14 - 6	15 - 7	16 - 8	17 - 9
9 - 0	10 - 1	11 - 2	12 - 3	13 - 4	14 - 5	15 - 6	16 - 7	17 - 8	18 - 9

Mental Math

These problems are a little different than the ones we have been doing. They require a student to add or subtract 10 from another two-digit number. If students understand place value, they should be able to take one from the tens place mentally to find the answer. You many want to start by writing each problem and solving it with the blocks. When the student is comfortable with this process, move on to doing the problems orally.

1. Ten plus ten equals what number? (20)
2. Sixteen plus ten equals what number? (26)
3. Twenty-five plus ten equals what number? (35)
4. Forty-two plus ten equals what number? (52)
5. Sixty-one plus ten equals what number? (71)
6. Thirty minus ten equals what number? (20)
7. Seventy-five minus ten equals what number? (65)
8. Twenty-one minus ten equals what number? (11)
9. Fifty-six minus ten equals what number? (46)
10. One hundred minus ten equals what number? (90)

LESSON 28

Subtraction by 6, or Adding Up by 4

There are four facts to be learned here. They are 11 – 6, 13 – 6, 14 – 6, and 15 – 6. As the title suggests, we could call these either subtracting by six or adding up by four. Study these facts carefully and add them to the ones you have already learned. Be sure to use the blocks and think about what you are doing.

Example 1

$$\begin{array}{r} 11 \\ -\ 6 \\ \hline \end{array}$$

10 plus what equals 11?
The answer is 1.
6 plus what equals 10?
The answer is 4.

$$\begin{array}{r} 11 \\ 10 \\ -\ 6 \\ \hline 5 \end{array}$$

(11 to 10: 1; 10 to 6: 4)

The difference between 6 and 11 is 4 + 1, or 5.

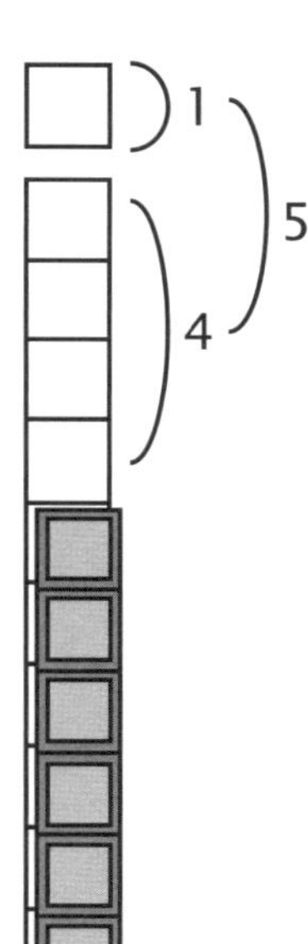

Example 2

$$\begin{array}{r} 13 \\ -\ 6 \\ \hline \end{array}$$

$$\begin{array}{r} 13 \\ 10 \\ -\ 6 \\ \hline 7 \end{array} \quad \begin{array}{l} 3 \\ 4 \end{array}$$

10 plus what equals 13?
The answer is 3.
6 plus what equals 10?
The answer is 4.

The difference between 6 and 13 is 4 + 3, or 7.

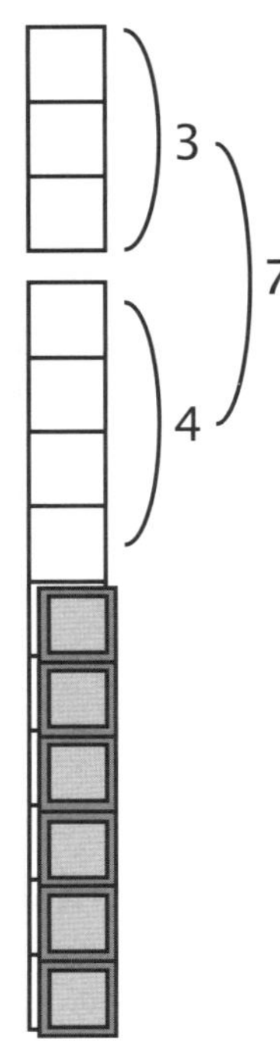

Example 3

$$\begin{array}{r} 15 \\ -\ 6 \\ \hline \end{array}$$

$$\begin{array}{r} 15 \\ 10 \\ -\ 6 \\ \hline 9 \end{array} \quad \begin{array}{l} 5 \\ 4 \end{array}$$

10 plus what equals 15?
The answer is 5.
6 plus what equals 10?
The answer is 4.

The difference between 6 and 15 is 4 + 5, or 9.

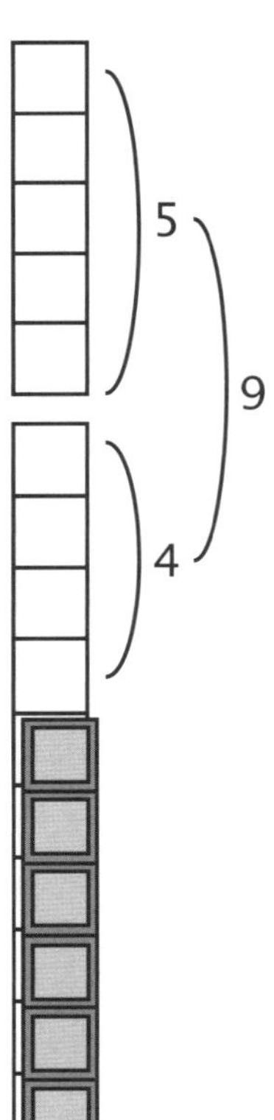

These four new facts are in the minus six column. Ninety-one facts have been mastered; we are counting down to the final nine.

0 - 0	1 - 1	2 - 2	3 - 3	4 - 4	5 - 5	6 - 6	7 - 7	8 - 8	9 - 9
1 - 0	2 - 1	3 - 2	4 - 3	5 - 4	6 - 5	7 - 6	8 - 7	9 - 8	10 - 9
2 - 0	3 - 1	4 - 2	5 - 3	6 - 4	7 - 5	8 - 6	9 - 7	10 - 8	11 - 9
3 - 0	4 - 1	5 - 2	6 - 3	7 - 4	8 - 5	9 - 6	10 - 7	11 - 8	12 - 9
4 - 0	5 - 1	6 - 2	7 - 3	8 - 4	9 - 5	10 - 6	11 - 7	12 - 8	13 - 9
5 - 0	6 - 1	7 - 2	8 - 3	9 - 4	10 - 5	11 - 6	12 - 7	13 - 8	14 - 9
6 - 0	7 - 1	8 - 2	9 - 3	10 - 4	11 - 5	12 - 6	13 - 7	14 - 8	15 - 9
7 - 0	8 - 1	9 - 2	10 - 3	11 - 4	12 - 5	13 - 6	14 - 7	15 - 8	16 - 9
8 - 0	9 - 1	10 - 2	11 - 3	12 - 4	13 - 5	14 - 6	15 - 7	16 - 8	17 - 9
9 - 0	10 - 1	11 - 2	12 - 3	13 - 4	14 - 5	15 - 6	16 - 7	17 - 8	18 - 9

LESSON 29

Subtraction by 5, or Adding Up by 5

The four new facts in this lesson are 11 – 5, 12 – 5, 13 – 5, and 14 – 5. As you've come to expect, we can refer to them as subtracting by five or as adding up by five. Have fun memorizing these facts.

Example 1

$$\begin{array}{r} 12 \\ -\ 5 \\ \hline \end{array}$$

$$\begin{array}{r} 12 \\ 10 \\ -\ 5 \\ \hline 7 \end{array}$$

(12 to 10: 2; 10 to 5: 5)

10 plus what equals 12?
The answer is 2.
5 plus what equals 10?
The answer is 5.

The difference between 5 and 12 is 5 + 2, or 7.

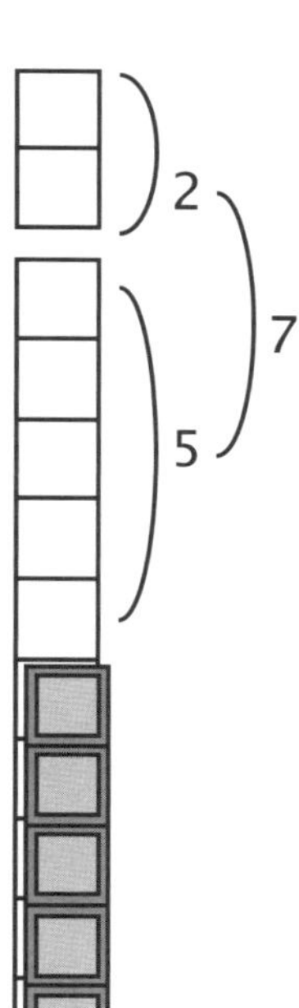

Example 2

$$\begin{array}{r} 13 \\ -\ 5 \\ \hline \end{array}$$

10 plus what equals 13?
The answer is 3.
5 plus what equals 10?
The answer is 5.

$$\begin{array}{r} 13 \\ 10 \\ -\ 5 \\ \hline 8 \end{array}$$

(13 to 10: 3; 10 to 5: 5)

The difference between 5 and 13 is 5 + 3, or 8.

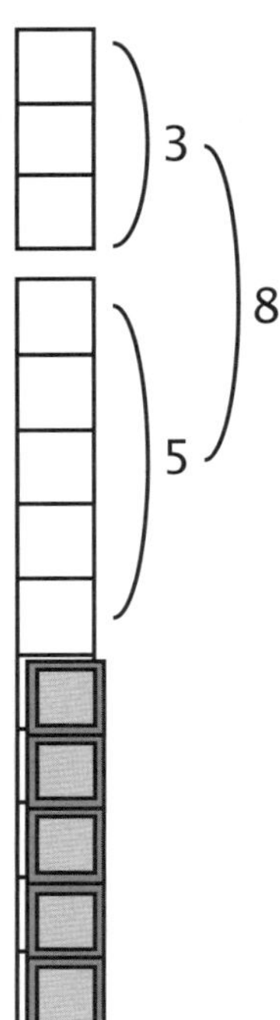

Example 3

$$\begin{array}{r} 14 \\ -\ 5 \\ \hline \end{array}$$

10 plus what equals 14?
The answer is 4.
5 plus what equals 10?
The answer is 5.

$$\begin{array}{r} 14 \\ 10 \\ -\ 5 \\ \hline 9 \end{array}$$

(14 to 10: 4; 10 to 5: 5)

The difference between 5 and 14 is 5 + 4, or 9.

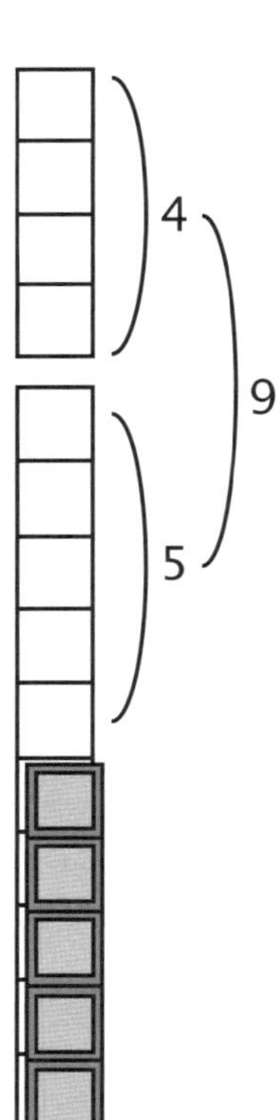

After memorizing these four new facts, we have mastered 95 facts and are down to the final five. The new facts appear in the minus five column.

0 – 0	1 – 1	2 – 2	3 – 3	4 – 4	5 – 5	6 – 6	7 – 7	8 – 8	9 – 9
1 – 0	2 – 1	3 – 2	4 – 3	5 – 4	6 – 5	7 – 6	8 – 7	9 – 8	10 – 9
2 – 0	3 – 1	4 – 2	5 – 3	6 – 4	7 – 5	8 – 6	9 – 7	10 – 8	11 – 9
3 – 0	4 – 1	5 – 2	6 – 3	7 – 4	8 – 5	9 – 6	10 – 7	11 – 8	12 – 9
4 – 0	5 – 1	6 – 2	7 – 3	8 – 4	9 – 5	10 – 6	11 – 7	12 – 8	13 – 9
5 – 0	6 – 1	7 – 2	8 – 3	9 – 4	10 – 5	11 – 6	12 – 7	13 – 8	14 – 9
6 – 0	7 – 1	8 – 2	9 – 3	10 – 4	11 – 5	12 – 6	13 – 7	14 – 8	15 – 9
7 – 0	8 – 1	9 – 2	10 – 3	11 – 4	12 – 5	13 – 6	14 – 7	15 – 8	16 – 9
8 – 0	9 – 1	10 – 2	11 – 3	12 – 4	13 – 5	14 – 6	15 – 7	16 – 8	17 – 9
9 – 0	10 – 1	11 – 2	12 – 3	13 – 4	14 – 5	15 – 6	16 – 7	17 – 8	18 – 9

More Solving for the Unknown

In lesson 8, students learned to solve for the unknown when one part of an addition problem was missing. Some of the games and activities have helped students explore different ways to make a number from other numbers. We have also used the skill of counting up to solve subtraction problems.

All of these skills can be used to help the student solve word problems with the unknown numbers in different positions. Here are some examples. Do these together and use the blocks as much as necessary.

1. Cassie broke three of her balloons. She had five balloons left. How many balloons did Cassie have before any were broken?

   B – 3 = 5, so B must be 8.

2. Bill had 6 balloons. After he broke some of the balloons, he had two left. How many balloons did Bill break? 6 – B = 2, so B must be 4.

LESSON 30

Subtraction by 3 and 4

There are only five more facts, and we are done. These seem more difficult than most, but they are not too hard when we know how to add. The facts are 11 – 3, 12 – 3, 11 – 4, 12 – 4, and 13 – 4. I'm still convinced that adding up is the clearest way to learn these facts, but you can also simply memorize them.

Example 1

$$\begin{array}{r} 12 \\ -\ 4 \\ \hline \end{array}$$

10 plus what equals 12?
The answer is 2.
4 plus what equals 10?
The answer is 6.

$$\begin{array}{r} 12 \\ 10 \\ -\ 4 \\ \hline 8 \end{array}$$

(12 to 10: 2; 10 to 4: 6)

The difference between 4 and 12 is 6 + 2, or 8.

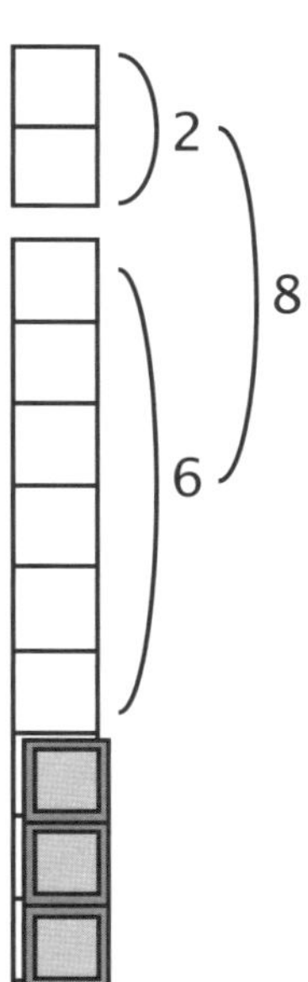

Example 2

$$\begin{array}{r} 13 \\ -\ 4 \\ \hline \end{array}$$

10 plus what equals 13?
The answer is 3.
4 plus what equals 10?
The answer is 6.

$$\begin{array}{r} 13 \\ 10 \\ -\ 4 \\ \hline 9 \end{array} \quad \begin{array}{l} 3 \\ 6 \end{array}$$

The difference between 4 and 13 is 6 + 3, or 9.

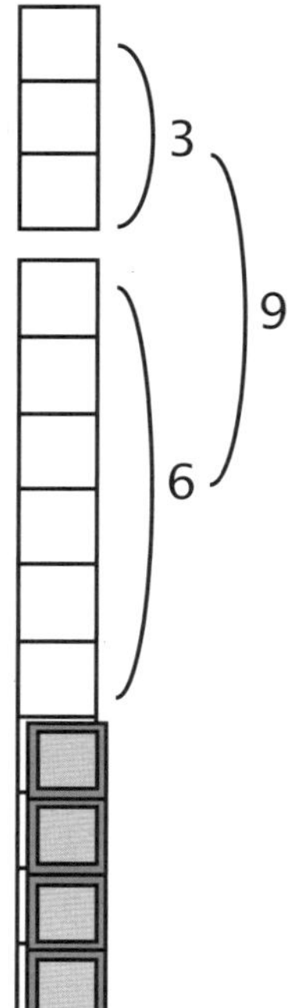

Example 3

$$\begin{array}{r} 14 \\ -\ 5 \\ \hline \end{array}$$

10 plus what equals 14?
The answer is 4.
5 plus what equals 10?
The answer is 5.

$$\begin{array}{r} 14 \\ 10 \\ -\ 5 \\ \hline 9 \end{array} \quad \begin{array}{l} 4 \\ 5 \end{array}$$

The difference between 5 and 14 is 5 + 4, or 9.

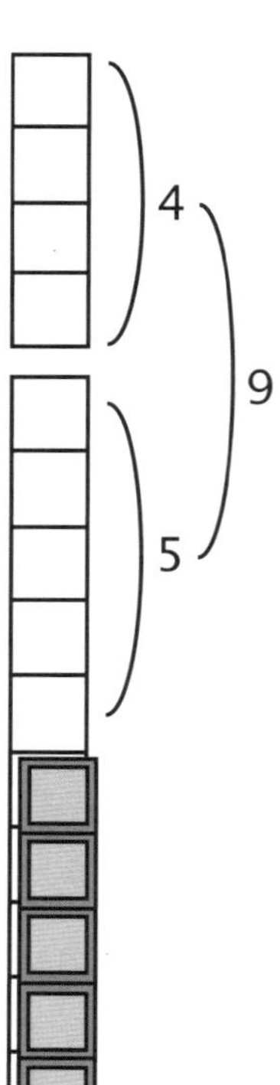

You made it! What a whiz you are! Congratulations! All 100 subtraction facts have been learned.

0 - 0	1 - 1	2 - 2	3 - 3	4 - 4	5 - 5	6 - 6	7 - 7	8 - 8	9 - 9
1 - 0	2 - 1	3 - 2	4 - 3	5 - 4	6 - 5	7 - 6	8 - 7	9 - 8	10 - 9
2 - 0	3 - 1	4 - 2	5 - 3	6 - 4	7 - 5	8 - 6	9 - 7	10 - 8	11 - 9
3 - 0	4 - 1	5 - 2	6 - 3	7 - 4	8 - 5	9 - 6	10 - 7	11 - 8	12 - 9
4 - 0	5 - 1	6 - 2	7 - 3	8 - 4	9 - 5	10 - 6	11 - 7	12 - 8	13 - 9
5 - 0	6 - 1	7 - 2	8 - 3	9 - 4	10 - 5	11 - 6	12 - 7	13 - 8	14 - 9
6 - 0	7 - 1	8 - 2	9 - 3	10 - 4	11 - 5	12 - 6	13 - 7	14 - 8	15 - 9
7 - 0	8 - 1	9 - 2	10 - 3	11 - 4	12 - 5	13 - 6	14 - 7	15 - 8	16 - 9
8 - 0	9 - 1	10 - 2	11 - 3	12 - 4	13 - 5	14 - 6	15 - 7	16 - 8	17 - 9
9 - 0	10 - 1	11 - 2	12 - 3	13 - 4	14 - 5	15 - 6	16 - 7	17 - 8	18 - 9

APPENDIX A

Telling Time: Minutes

When the student has mastered skip counting by fives, he is ready to learn how to tell time. This can be challenging with a clock that is not digital. We'll begin by taking six 10 bars and explaining that there are 60 minutes in one hour. Next replace each 10 bar with two five bars. If you don't have 12 five bars, use five units, a four and one bar, or a three and two bar. Arrange your 12 groups of five in a circle (really a dodecagon, or 12-sided polygon) so that your 60 minutes are in the shape of a clock. Starting at the top, begin skip counting by fives and go around the clock: 5–10–15–20–25–30–35–40–45–50–55–60.

Choose any long bar, turn it on its side so that it is smooth, and use it as your minute hand. Point it at different areas and, beginning at the top, count the minutes. Using one set of blocks, you can make this clock by using various combinations for fives (five unit pieces, a two and a three, or a four and a unit). There is a removable clock template for your use at the end of the student book.

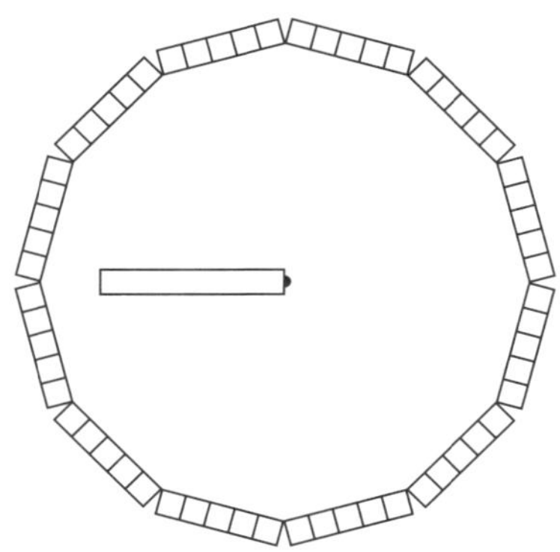

To help the students see the progression of the minutes, build several partial clocks as in Examples 1 and 2. Count the minutes by beginning at the top and moving around to the right, or clockwise.

Example 1

Count the minutes: 5-10, so 10 minutes

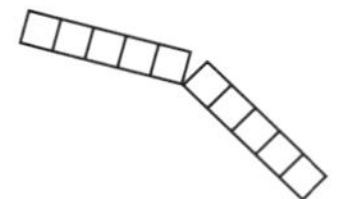

Example 2

Count the minutes: 5-10-15, so 15 minutes

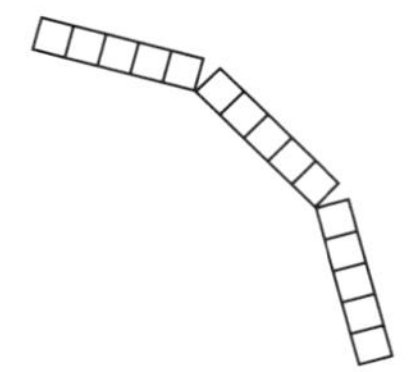

APPENDIX B

Telling Time: Hours

Once the minutes are mastered, we can add the hours by placing a green unit bar at the end of the first five bar (outside the circle), pointing away from the center of the clock. (See the diagram on the next page.) To distinguish between the minutes and hours, I leave the minutes bars right side up but place the hours upside down with the hollow side showing. You can still see the color and how many hours there are, but it helps to distinguish the minutes from the hours. Place your orange bar (upside down so the hollow side is showing) at the end of the second five bar. Continue this process with all the unit bars through 12. Choose a unit bar smaller than your minute hand for an hour hand. Turn it upside down so the student makes the connection between the hour hand and the hours, since both are upside down with the hollow side showing.

Position the hour hand so that it points between the two and the three. This is the critical point for telling time. Is it two o'clock or three o'clock? I've explained this to many children with success by aiming the hour hand at the two and saying, "He just had his second birthday and is now two." Then I move the hand towards the three and ask, "How old is he now? Is he still two?" "Yes." Then I move it a little farther and ask if he's still two. "Yes." I do this until the hand is almost pointing to three and ask the question, "What about the day before his next birthday; how old is he?" "Still two." He is almost three but still only two. Practice this skill until the student can confidently identify the hour by moving the hour hand around the clock on his own.

When the hour hand is mastered, put the hours and minutes together. Have the student first identify the hour and then skip count to find the minutes. Finally, when this is mastered, have the student tell time without the manipulatives by looking at a real clock face.

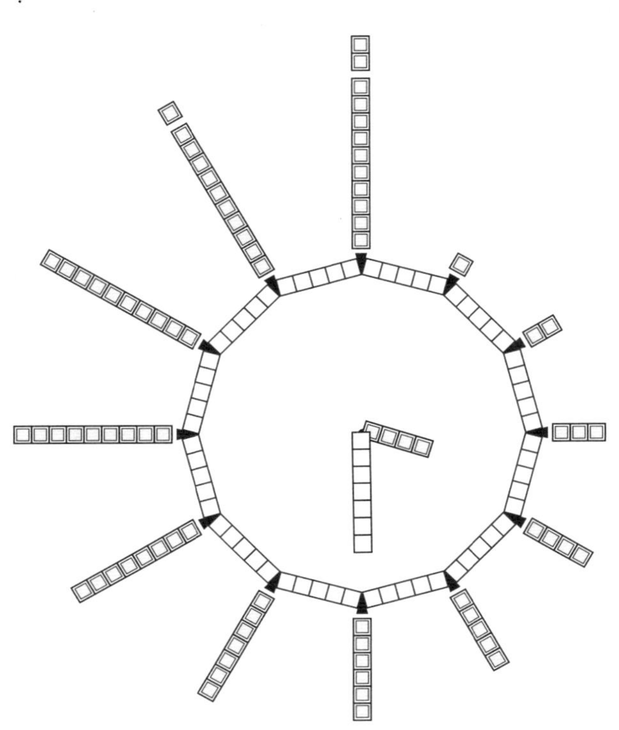

Student Solutions

Lesson Practice 1A

1. color 3 hundreds, 2 tens, and 4 units; three hundred twenty-four
2. 172; one hundred seventy-two
3. 3 tens and 4 units; thirty-four
4. 2 hundreds, 3 tens, and 5 units; two hundred thirty-five

Lesson Practice 1B

1. color 6 hundreds, 4 tens and 7 units; six hundred forty-seven
2. 452; four hundred fifty-two
3. 1 ten and 6 units; sixteen
4. 1 hundred, 9 tens, and 8 units; one hundred ninety-eight

Lesson Practice 1C

1. color 8 hundreds, 1 ten, and 3 units; eight hundred thirteen
2. 125; one hundred twenty-five
3. 6 tens and 7 units; sixty-seven
4. 3 hundreds, 2 tens, and 6 units; three hundred twenty-six

Lesson Practice 1D

1. color 4 hundreds, 9 tens, and 0 units four hundred ninety
2. 591; five hundred ninety-one
3. 4 hundreds, 5 tens, and 1 unit; four hundred fifty-one
4. 1 ten and 4 units; fourteen

Lesson Practice 2A

1. 0, 1, 2, 3, 4, 5, 6, 7, 8, 9, 10, 11, 12, 13, 14, 15, 16, 17, 18, 19, 20
2. 0, 1, 2, 3, 4, 5, 6, 7, 8, 9, 10, 11, 12, 13, 14, 15, 16, 17, 18, 19, 20
3. 0, 1, 2, 3, 4, 5, 6, 7, 8, 9, 10, 11, 12, 13, 14, 15, 16, 17, 18, 19, 20
4. 0, 1, 2, 3, 4, 5, 6, 7, 8, 9, 10, 11, 12, 13, 14, 15, 16, 17, 18, 19, 20

Lesson Practice 2B

1. 0, 1, 2, 3, 4, 5, 6, 7, 8, 9, 10, 11, 12, 13, 14, 15, 16, 17, 18, 19, 20
2. 0, 1, 2, 3, 4, 5, 6, 7, 8, 9, 10, 11, 12, 13, 14, 15, 16, 17, 18, 19, 20
3. 0, 1, 2, 3, 4, 5, 6, 7, 8, 9, 10, 11, 12, 13, 14, 15, 16, 17, 18, 19, 20
4. 0, 1, 2, 3, 4, 5, 6, 7, 8, 9, 10, 11, 12, 13, 14, 15, 16, 17, 18, 19, 20

Lesson Practice 2C

1. 0, 1, 2, 3, 4, 5, 6, 7, 8, 9, 10, 11, 12, 13, 14, 15, 16, 17, 18, 19, 20
2. 0, 1, 2, 3, 4, 5, 6, 7, 8, 9, 10, 11, 12, 13, 14, 15, 16, 17, 18, 19, 20
3. 0, 1, 2, 3, 4, 5, 6, 7, 8, 9, 10, 11, 12, 13, 14, 15, 16, 17, 18, 19, 20
4. 0, 1, 2, 3, 4, 5, 6, 7, 8, 9, 10, 11, 12, 13, 14, 15, 16, 17, 18, 19, 20

Systematic Review 2D

1. 0, 1, 2, 3, 4, 5, 6, 7, 8, 9, 10, 11, 12, 13, 14, 15, 16, 17, 18, 19, 20
2. 462; four hundred sixty-two
3. two hundred forty-two
4. seventeen

Systematic Review 2E

1. 0, 1, 2, 3, 4, 5, 6, 7, 8, 9, 10, 11, 12, 13, 14, 15, 16, 17, 18, 19, 20
2. four hundred eighteen
3. fifty-five
4. three hundred seventy-nine

Systematic Review 2F

1. 0, 1, 2, 3, 4, 5, 6, 7, 8, 9, 10, 11, 12, 13, 14, 15, 16, 17, 18, 19, 20
2. 0, 1, 2, 3, 4, 5, 6, 7, 8, 9, 10, 11, 12, 13, 14, 15, 16, 17, 18, 19, 20
3. 146; one hundred forty-six
4. two hundred three
5. eighty-one

Lesson Practice 3A

1. 5 orange
   3 pink
   2 light blue
2. 4 yellow
   6 light green
   9 violet
3. brown 8-block; eight
4. violet 6-block; six
5. tan
6. orange

Lesson Practice 3B

1. 1 tan
   8 dark green
   7 brown
2. 3 light blue
   9 pink
   5 light green
3. dark blue 10-block; ten
4. yellow 4-block; four
5. violet
6. dark green

Lesson Practice 3C

1. 4 tan
   6 violet
   7 yellow
2. 9 brown
   5 light blue
   8 light green
3. orange 2-block; two
4. pink 3-block; three
5. yellow
6. brown

Systematic Review 3D

1. 2 light blue
   5 pink
   3 orange
   1 green
2. violet 6-block; six
3. light green
4. 203; two hundred three
5. 0, 1, 2, 3, 4, 5, 6, 7, 8, 9, 10, 11, 12, 13, 14, 15, 16, 17, 18, 19, 20

Systematic Review 3E

1. 4 light green
   8 yellow
   9 brown
2. light blue 5-block; five
3. orange
4. 1 hundred, 3 tens, and 5 units; one hundred thirty-five
5. 0, 1, 2, 3, 4, 5, 6, 7, 8, 9, 10, 11, 12, 13, 14, 15, 16, 17, 18, 19, 20

Systematic Review 3F

1. 8 pink
   6 light blue
   3 violet
   5 brown
2. yellow 4-block; four
3. pink
4. 3 hundreds, 5 tens, and 7 units; three hundred fifty-seven
5. 0, 1, 2, 3, 4, 5, 6, 7, 8, 9, 10, 11, 12, 13, 14, 15, 16, 17, 18, 19, 20

Lesson Practice 4A

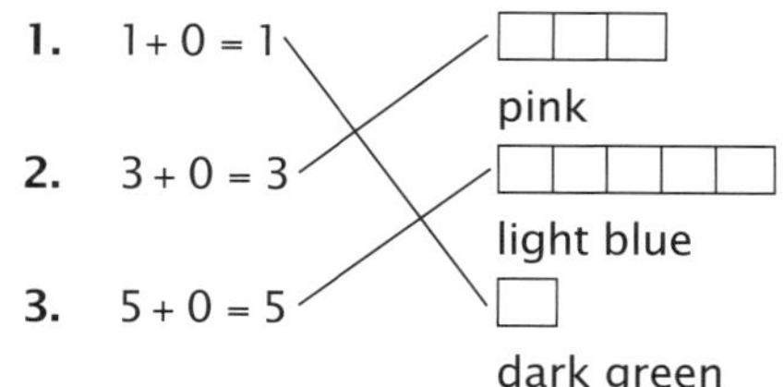

5. 7 + 0 = 7
6. 9 + 0 = 9
7. 4 + 0 = 4
8. 6 + 0 = 6
9. 5 + 0 = 5
10. 0 + 3 = 3
11. 0 + 2 = 2
12. 0 + 0 = 0
13. done
14. 1 + 0 = 1 time

Lesson Practice 4B

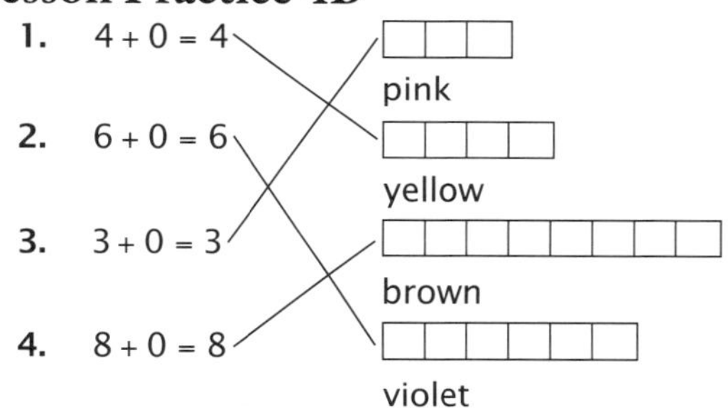

5. 5 + 0 = 5
6. 0 + 0 = 0
7. 7 + 0 = 7
8. 1 + 0 = 1
9. 3 + 0 = 3
10. 0 + 6 = 6
11. 0 + 9 = 9
12. 0 + 2 = 2
13. 4 + 0 = 4 stories
14. 0 + 8 = 8 pets

Lesson Practice 4C

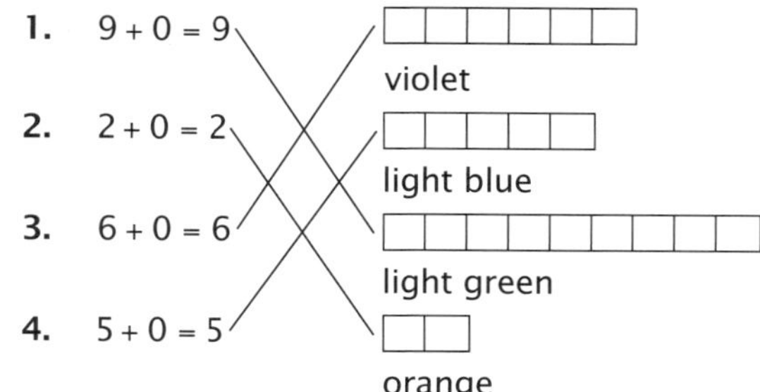

5. 0 + 1 = 1
6. 6 + 0 = 6
7. 9 + 0 = 9
8. 2 + 0 = 2
9. 0 + 0 = 0
10. 0 + 5 = 5
11. 0 + 7 = 7
12. 8 + 0 = 8
13. 7 + 0 = 7 dollars
14. 0 + 0 = 0 pictures

Systematic Review 4D

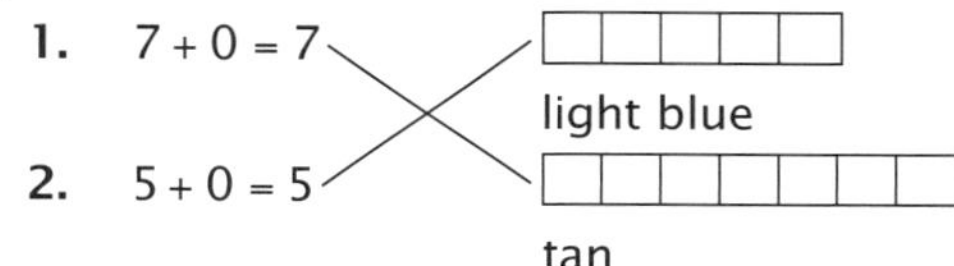

3. 1 + 0 = 1
4. 8 + 0 = 8
5. 0 + 6 = 6
6. 0 + 0 = 0
7. 9 + 0 = 9
8. 0 + 4 = 4
9. 0 + 3 = 3
10. 2 + 0 = 2
11. 0, 1, 2, 3, 4, 5, 6, 7, 8, 9, 10, 11, 12, 13, 14, 15, 16, 17, 18, 19, 20
12. 1 hundred, 8 tens, and 2 units; one hundred eighty-two
13. 2 hundreds and 9 units
14. tan
15. 3 + 0 = 3 helpings

Systematic Review 4E

1. 2 + 0 = 2 — orange
2. 4 + 0 = 4 — yellow
3. 8 + 0 = 8
4. 5 + 0 = 5
5. 0 + 0 = 0
6. 9 + 0 = 9
7. 1 + 0 = 1
8. 0 + 6 = 6
9. 7 + 0 = 7
10. 0 + 3 = 3
11. 0, 1, 2, 3, 4, 5, 6, 7, 8, 9, 10, 11, 12, 13, 14, 15, 16, 17, 18, 19, 20
12. 3 hundreds, 1 ten, and 4 units; three hundred fourteen
13. 4 hundreds, and 2 tens; four hundred twenty
14. brown
15. 0 + 2 = 2 inches

Systematic Review 4F

1. 6 + 0 = 6
2. 9 + 0 = 9

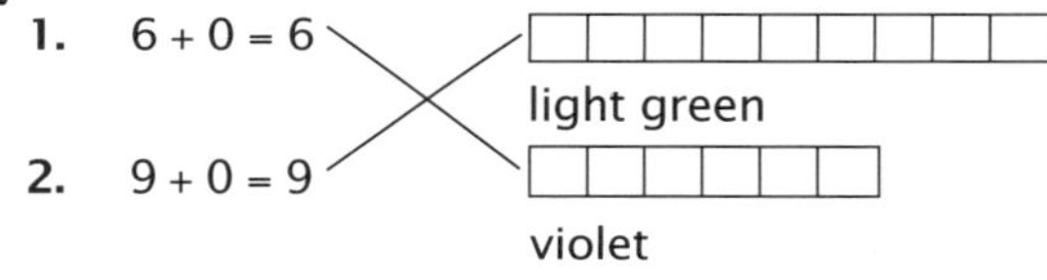

light green

violet

3. 1 + 0 = 1
4. 6 + 0 = 6
5. 0 + 2 = 2
6. 4 + 0 = 4
7. 7 + 0 = 7
8. 0 + 0 = 0
9. 3 + 0 = 3
10. 0 + 5 = 5
11. 0, 1, 2, 3, 4, 5, 6, 7, 8, 9, 10, 11, 12, 13, 14, 15, 16, 17, 18, 19, 20
12. 1 hundred, 4 tens, and 8 units; one hundred forty-eight
13. 3 hundreds, 2 tens, and 5 units; three hundred twenty-five
14. light blue
15. 0 + 6 = 6 cookies

Lesson Practice 5A

1. 2 + 1 = 3
2. 3 + 1 = 4
3. 5 + 1 = 6
4. 1 + 1 = 2

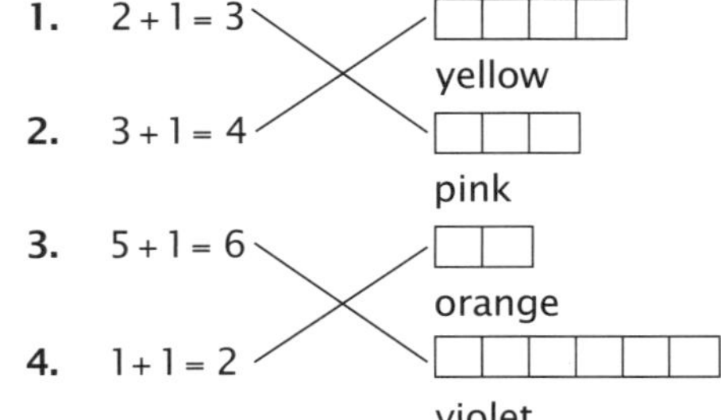

yellow

pink

orange

violet

5. 4 + 1 = 5
6. 6 + 1 = 7
7. 8 + 1 = 9
8. 7 + 1 = 8
9. 2 + 1 = 3
10. 1 + 5 = 6
11. 1 + 3 = 4
12. 1 + 1 = 2
13. done
14. 1 + 4 = 5 flowers

Lesson Practice 5B

1. 7 + 1 = 8
2. 5 + 1 = 6
3. 9 + 1 = 10
4. 8 + 1 = 9

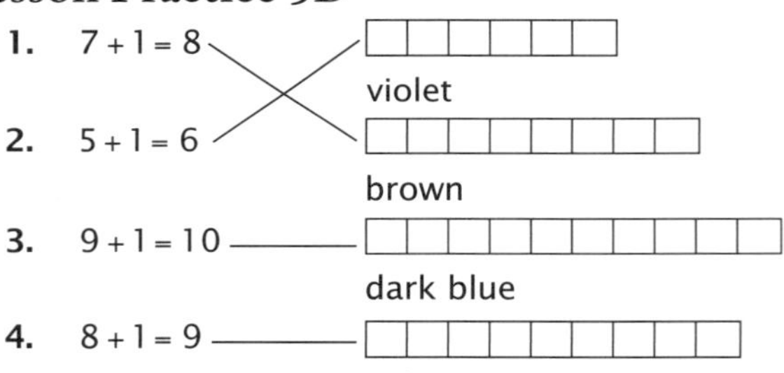

violet

brown

dark blue

light green

5. 1 + 1 = 2
6. 3 + 1 = 4
7. 5 + 1 = 6
8. 2 + 1 = 3
9. 7 + 1 = 8
10. 1 + 8 = 9
11. 1 + 6 = 7
12. 4 + 1 = 5
13. 5 + 1 = 6 CDs
14. 1 + 9 = 10 pots

Lesson Practice 5C

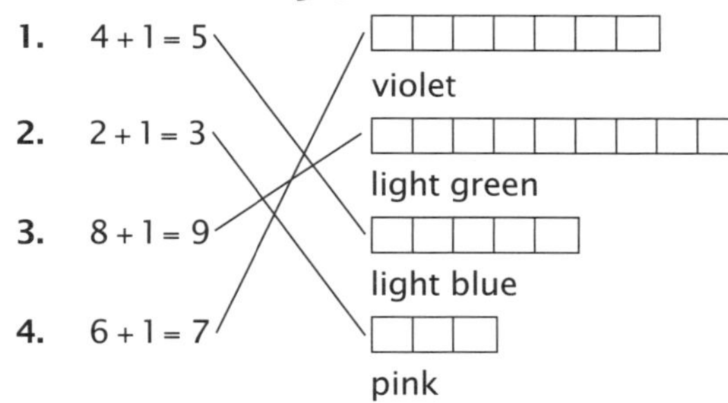

5. 1 + 0 = 1
6. 9 + 1 = 10
7. 1 + 5 = 6
8. 3 + 1 = 4
9. 1 + 1 = 2
10. 1 + 7 = 8
11. 8 + 1 = 9
12. 6 + 1 = 7
13. 2 + 1 = 3 candy canes
14. 4 + 1 = 5 tires

Systematic Review 5D

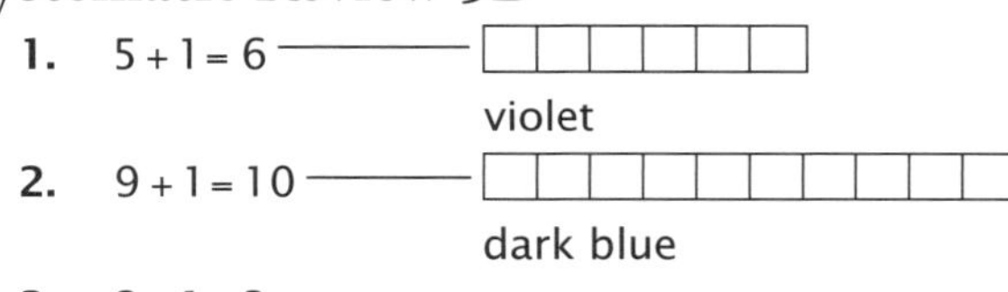

3. 2 + 1 = 3
4. 7 + 0 = 7
5. 1 + 9 = 10
6. 7 + 1 = 8
7. 4 + 0 = 4
8. 1 + 5 = 6
9. 6 + 1 = 7
10. 8 + 0 = 8
11. 3 + 1 = 4
12. 1 + 8 = 9
13. 4 + 1 = 5
14. 1 + 0 = 1
15. 315; three hundred fifteen
16. 1 hundred and 5 tens;
    one hundred fifty
17. 4 hundreds, 7 tens, and 3 units;
    four hundred seventy-three
18. 3 + 1 = 4 cookies
    4 + 0 = 4 cookies

Systematic Review 5E

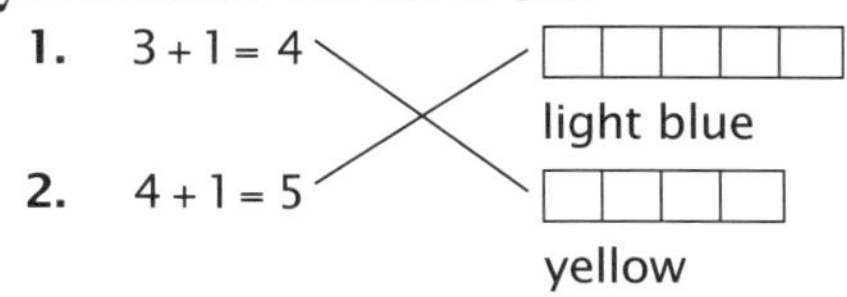

3. 1 + 6 = 7
4. 8 + 1 = 9
5. 0 + 4 = 4
6. 1 + 5 = 6
7. 3 + 0 = 3
8. 1 + 7 = 8
9. 9 + 1 = 10
10. 1 + 1 = 2
11. 2 + 1 = 3
12. 8 + 0 = 8
13. 3 + 1 = 4
14. 1 + 4 = 5
15. 172; one hundred seventy-two
16. 3 hundreds and 9 units;
    three hundred nine
17. 8 tens and 5 units;
    eighty-five
18. 7 + 0 = 7 dollars

Systematic Review 5F

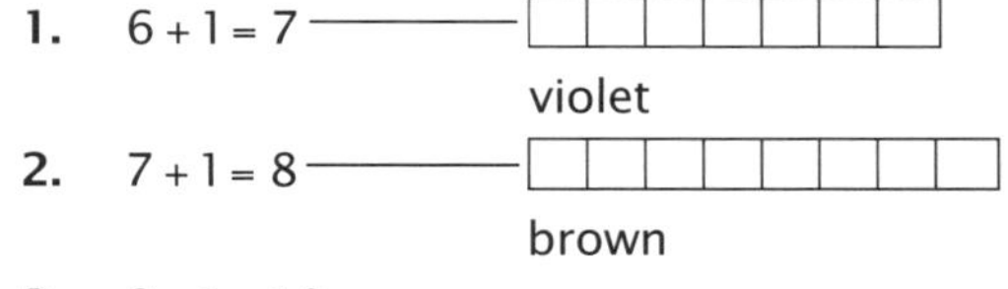

3. 9 + 1 = 10
4. 3 + 0 = 3
5. 1 + 8 = 9
6. 6 + 1 = 7
7. 1 + 1 = 2
8. 0 + 5 = 5
9. 6 + 0 = 6
10. 4 + 1 = 5

11. $2 + 1 = 3$
12. $1 + 5 = 6$
13. $3 + 1 = 4$
14. $1 + 7 = 8$
15. 223; two hundred twenty-three
16. 4 hundreds, 1 ten, and 8 units; four hundred eighteen
17. 2 hundreds and 1 unit; two hundred one
18. $3 + 1 = 4$ inches
    $4 + 1 = 5$ inches

Lesson Practice 6A

1.

0	1	2	3	4	5	6	7	8	9
10	11	12	13	14	15	16	17	18	19
20	21	22	23	24	25	26	27	28	29
30	31	32	33	34	35	36	37	38	39
40	41	42	43	44	45	46	47	48	49
50	51	52	53	54	55	56	57	58	59
60	61	62	63	64	65	66	67	68	69
70	71	72	73	74	75	76	77	78	79
80	81	82	83	84	85	86	87	88	89
90	91	92	93	94	95	96	97	98	99
100									

2. 10, 20, 30, 40, 50, 60

Lesson Practice 6B

1. see grid for 6A – 1
2. 10, 20, 30, 40, 50, 60, 70, 80

Lesson Practice 6C

1. see grid for 6A – 1
2. 10, 20, 30, 40, 50, 60, 70, 80, 90, 100

Systematic Review 6D

1. see grid for 6A – 1
2. 10, 20, 30, 40, 50, 60, 70, 80, 90, 100
3. $1 + 1 = 2$
4. $8 + 0 = 8$
5. $5 + 1 = 6$
6. $7 + 0 = 7$

Systematic Review 6E

1. see grid for 6A – 1
2. 10, 20, 30, 40, 50, 60, 70, 80, 90, 100
3. $0 + 5 = 5$
4. $6 + 1 = 7$
5. $4 + 1 = 5$
6. $0 + 3 = 3$

Systematic Review 6F

1. see grid for 6A – 1
2. 10, 20, 30, 40, 50, 60, 70, 80, 90, 100
3. $1 + 2 = 3$
4. $7 + 1 = 8$
5. $9 + 1 = 10$
6. $0 + 4 = 4$

Lesson Practice 7A

1. green (1) + orange(2) = pink (3)
2. tan (7) + orange(2) = light green (9)
3. pink (3) + orange(2) = light blue (5)
4. orange(2) + orange(2) = yellow (4)
5. $6 + 2 = 8$
6. $8 + 2 = 10$
7. $50 + 20 = 70$
8. $200 + 200 = 400$
9. $7 + 2 = 9$
10. $4 + 2 = 6$
11. $2 + 3 = 5$
12. $0 + 2 = 2$
13. done
14. $8 + 2 = 10$ cars

Lesson Practice 7B

1. light blue(5) + orange(2) = violet (7)
2. yellow(4) + orange(2) = light blue (6)

3. brown(8) + orange(2) = dark blue (10)
4. (0) + orange(2) = orange (2)
5. 2 + 2 = 4
6. 7 + 2 = 9
7. 20 + 60 = 80
8. 100 + 200 = 300
9. 2 + 5 = 7
10. 4 + 2 = 6
11. 2 + 8 = 10
12. 3 + 2 = 5
13. 7 + 2 = 9 sailboats
14. 4 + 2 = 6 crackers

Lesson Practice 7C

1. violet(6) + orange(2) = dark brown (8)

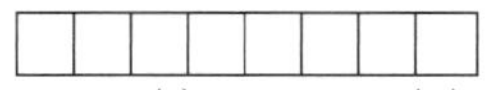

2. green(1) + orange(2) = pink (3)
3. tan(7) + orange(2) = light green (9)
4. pink(3) + orange(2) = light blue (5)
5. 2 + 8 = 10
6. 0 + 2 = 2
7. 40 + 20 = 60
8. 200 + 100 = 300
9. 2 + 2 = 4
10. 2 + 5 = 7
11. 7 + 2 = 9
12. 3 + 2 = 5
13. 5 + 2 = 7 lollipops
14. 3 + 2 = 5 children

Systematic Review 7D

1. brown(8) + orange(2) = dark blue (10)
2. light blue(5) + orange(2) = tan (7)
3. 3 + 2 = 5
4. 2 + 6 = 8
5. 2 + 4 = 6
6. 70 + 20 = 90
7. 300 + 100 = 400
8. 9 + 0 = 9
9. 1 + 2 = 3
10. 1 + 8 = 9
11. 2 + 0 = 2
12. 1 + 6 = 7
13. 9 + 1 = 10
14. 1 + 3 = 4
15. 2 hundreds, 6 tens, and 4 units;
two hundred sixty-four
16. 10, 20, 30, 40, 50, 60, 70, 80, 90, 100
17. 2 + 2 = 4 games
18. 6 + 1 = 7 pennies

Systematic Review 7E

1. orange(2) + orange(2) = yellow (4)

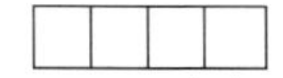

2. yellow(4) + orange(2) = violet (6)

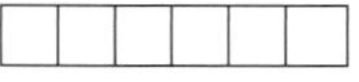

3. 2 + 5 = 7
4. 2 + 7 = 9
5. 30 + 20 = 50
6. 200 + 100 = 300
7. 6 + 2 = 8
8. 8 + 1 = 9
9. 8 + 2 = 10
10. 0 + 5 = 5
11. 0 + 6 = 6
12. 1 + 5 = 6
13. 3 + 0 = 3
14. 7 + 1 = 8
15. 4 hundreds and 7 units;
four hundred seven
16. 10, 20, 30, 40, 50, 60, 70, 80, 90, 100
17. 8 + 2 = 10 flowers
18. 7 + 2 = 9 points
9 + 1 = 10 points

Systematic Review 7F

1. pink (3) + orange (2) = light blue (5)

   $\square\square\square\square\square$

2. violet (6) + orange (2) = dark brown (8)

   $\square\square\square\square\square\square\square\square$

3. $2+4=6$
4. $7+2=9$
5. $50+20=70$
6. $1+2=3$
7. $2+0=2$
8. $2+2=4$
9. $8+2=10$
10. $100+100=200$
11. $9+0=9$
12. $7+0=7$
13. $9+1=10$
14. $1+3=4$
15. 3 hundreds, 2 tens, and 2 units;
    three hundred twenty-two
16. 8 tens and 1 unit;
    eighty-one
17. $2+4=6$ dollars
18. $5+2=7$ years old

Lesson Practice 8A

1. yellow(4) + orange(2) = violet(6)
2. yellow(4) + 0 = yellow(4)
3. tan(7) + orange(2) = light green(9)
4. dark green(1) + dark green(1) = orange(2)
5. $\boxed{3}+2=5$
6. $\boxed{5}+2=7$
7. $\boxed{8}+0=8$
8. $\boxed{8}+1=9$
9. $\boxed{6}+2=8$
10. $\boxed{3}+1=4$
11. done
12. $\boxed{1}+2=3$ places

Lesson Practice 8B

1. orange(2) + orange(2) = yellow(4)
2. violet(6) + 0 = violet(6)
3. violet(6) + orange(2) = brown(8)
4. light blue(5) + dark green(1) = violet(6)
5. $\boxed{8}+2=10$
6. $\boxed{6}+1=7$
7. $\boxed{9}+0=9$
8. $\boxed{7}+2=9$
9. $\boxed{4}+1=5$
10. $\boxed{0}+2=2$
11. $\boxed{8}+1=9$ children
12. $\boxed{2}+4=6$ items

Lesson Practice 8C

1. light blue(5) + 0 = light blue(5)
2. tan(7) + dark green(1) = dark brown(8)
3. dark brown(8) + dark green(1) =
   light green(9)
4. light blue(5) + orange(2) = tan(7)
5. $\boxed{1}+2=3$
6. $\boxed{9}+1=10$
7. $\boxed{2}+0=2$
8. $\boxed{4}+2=6$
9. $\boxed{0}+1=1$
10. $\boxed{3}+1=4$
11. $\boxed{3}+2=5$ tractors
12. $\boxed{7}+2=9$ people

Systematic Review 8D

1. $\boxed{0}+0=0$
2. $\boxed{8}+2=10$
3. $\boxed{2}+2=4$
4. $\boxed{6}+1=7$
5. $\boxed{4}+1=5$
6. $\boxed{3}+2=5$
7. $2+1=3$
8. $40+20=60$

9. $6+0=6$
10. $5+2=7$
11. $3+0=3$
12. $7+1=8$
13. $8+2=10$
14. $1+4=5$
15. 132; one hundred thirty-two
16. 69; sixty-nine
17. $2+0=2$ hands
18. $5+2=7$ loaves
    $7+\boxed{2}=9$ loaves

Systematic Review 8E

1. $\boxed{1}+2=3$
2. $\boxed{3}+2=5$
3. $\boxed{7}+0=7$
4. $\boxed{9}+1=10$
5. $\boxed{6}+2=8$
6. $\boxed{3}+1=4$
7. $1+0=1$
8. $60+10=70$
9. $1+5=6$
10. $7+2=9$
11. $2+2=4$
12. $2+6=8$
13. $4+0=4$
14. $1+3=4$
15. 124; one hundred twenty-four
16. 76; seventy-six
17. $5+1=6$ children
18. $4+1=5$ hats
    $5+2=7$ hats

Systematic Review 8F

1. $\boxed{0}+2=2$
2. $\boxed{4}+2=6$
3. $\boxed{7}+1=8$
4. $\boxed{5}+1=6$
5. $\boxed{1}+0=1$
6. $\boxed{5}+2=7$
7. $3+2=5$
8. $80+10=90$
9. $1+1=2$
10. $8+0=8$
11. $0+9=9$
12. $2+5=7$
13. $9+1=10$
14. $2+8=10$
15. 201; two hundred one
16. 30; thirty
17. $7+2=9$ tennis balls
18. $\boxed{2}+2=4$ people

Lesson Practice 9A

1. $9+8=$ $10+2=12$
2. $9+1=$ $10+3=13$
3. $9+3=$ $10+7=17$
4. $9+4=$ $10+0=10$
5. $9+9=18$
6. $9+5=14$
7. $9+2=11$
8. $9+3=12$
9. $9+6=15$
10. $1+9=10$
11. $9+4=13$
12. $9+8=17$
13. $9+7=16$ boys
14. $9+9=18$ books

Lesson Practice 9B

1. $9+2=$ $10+1=11$
2. $9+5=$ $10+6=16$
3. $9+7=$ $10+2=12$
4. $9+3=$ $10+4=14$
5. $9+8=17$
6. $9+4=13$
7. $9+7=16$
8. $9+1=10$
9. $5+9=14$
10. $0+9=9$

11. $9+3=12$
12. $9+9=18$
13. $9+6=15$ sets
14. $9+3=12$ CDs

Lesson Practice 9C

1. $9+6=$ — $10+5=15$
2. $9+9=$ — $10+8=18$
3. $9+4=$ — $10+3=13$
4. $9+7=$ — $10+6=16$
5. $7+9=16$
6. $9+8=17$
7. $9+2=11$
8. $9+4=13$
9. $6+9=15$
10. $9+3=12$
11. $9+9=18$
12. $9+0=9$
13. $9+5=14$ candies
14. $1+9=10$ dogs

Systematic Review 9D

1. $9+9=18$
2. $5+2=7$
3. $40+10=50$
4. $9+7=16$
5. $200+200=400$
6. $5+9=14$
7. $1+6=7$
8. $9+6=15$
9. $9+0=9$
10. $8+9=17$
11. $7+2=9$
12. $9+1=10$
13. $\boxed{9}+4=13$
14. $\boxed{4}+2=6$
15. 4 hundreds, 6 tens, and 1 unit;
    four hundred sixty-one
16. 10, 20, 30, 40, 50, 60, 70, 80, 90, 100
17. $8+9=17$ years old
18. $\boxed{1}+6=7$ guests

Systematic Review 9E

1. $9+3=12$
2. $4+9=13$
3. $60+20=80$
4. $0+4=4$
5. $2+9=11$
6. $9+9=18$
7. $8+2=10$
8. $300+100=400$
9. $5+9=14$
10. $2+4=6$
11. $9+5=14$
12. $1+7=8$
13. $\boxed{9}+8=17$
14. $\boxed{2}+5=7$
15. 2 hundreds, 4 tens, and 9 units;
    two hundred forty-nine
16. 10, 20, 30, 40, 50, 60, 70, 80, 90, 100
17. $\boxed{3}+9=12$ dollars
18. $4+2=6$ calls
    $6+9=15$ calls

Systematic Review 9F

1. $8+9=17$
2. $9+7=16$
3. $2+2=4$
4. $80+10=90$
5. $0+0=0$
6. $9+3=12$
7. $6+2=8$
8. $10+50=60$
9. $9+4=13$
10. $2+9=11$
11. $2+7=9$
12. $9+5=14$
13. $\boxed{6}+9=15$
14. $\boxed{2}+3=5$
15. 5 tens and 2 units; fifty-two
16. 10, 20, 30, 40, 50, 60, 70, 80, 90, 100
17. $\boxed{5}+2=7$ children

18. 5 + 2 = 7 ducks
7 + [2] = 9 ducks
The unknown may be put in either the first or the second blank of the equation.

Lesson Practice 10A

1. 8 + 5 =   10 + 1 = 11
2. 8 + 8 =   10 + 3 = 13
3. 8 + 3 =   10 + 4 = 14
4. 8 + 6 =   10 + 6 = 16
5. 8 + 1 = 9
6. 8 + 3 = 11
7. 8 + 7 = 15
8. 8 + 9 = 17
9. 8 + 2 = 10
10. 8 + 4 = 12
11. 8 + 5 = 13
12. 8 + 6 = 14
13. 8 + 2 = 10 snowballs
14. 8 + 4 = 12 goldfish

Lesson Practice 10B

1. 8 + 4 =   10 + 0 = 10
2. 8 + 9 =   10 + 5 = 15
3. 8 + 7 =   10 + 7 = 17
4. 8 + 2 =   10 + 2 = 12
5. 8 + 5 = 13
6. 8 + 8 = 16
7. 1 + 8 = 9
8. 8 + 3 = 11
9. 6 + 8 = 14
10. 8 + 7 = 15
11. 9 + 8 = 17
12. 8 + 4 = 12
13. 8 + 8 = 16 guests
14. 8 + 3 = 11 beach balls

Lesson Practice 10C

1. 8 + 1 =   10 + 1 = 11
2. 8 + 3 =   9
3. 8 + 6 =   10 + 6 = 16
4. 8 + 8 =   10 + 4 = 14
5. 9 + 8 = 17
6. 7 + 8 = 15
7. 8 + 0 = 8
8. 2 + 8 = 10
9. 8 + 5 = 13
10. 6 + 8 = 14
11. 4 + 8 = 12
12. 1 + 8 = 9
13. 8 + 6 = 14
14. 8 + 0 = 8

Systematic Review 10D

1. 8 + 2 = 10
2. 5 + 8 = 13
3. 8 + 7 = 15
4. 8 + 8 = 16
5. 9 + 5 = 14
6. 7 + 9 = 16
7. 20 + 40 = 60
8. 9 + 8 = 17
9. 1 + 7 = 8
10. 5 + 2 = 7
11. 3 + 0 = 3
12. 7 + 2 = 9
13. done
14. [4] + 8 = 12
15. 8 + 5 = 13 cents
16. 9 + 6 = 15 birds
17. [2] + 7 = 9
18. 8 + 1 = 9 apples
9 + 2 = 11 apples

Systematic Review 10E

1. 1 + 8 = 9
2. 8 + 7 = 15
3. 8 + 5 = 13
4. 6 + 8 = 14
5. 8 + 2 = 10
6. 9 + 6 = 15
7. 20 + 30 = 50
8. 3 + 9 = 12
9. 6 + 0 = 6

10. $2+2=4$
11. $9+4=13$
12. $6+2=8$
13. $\boxed{3}+8=11$
14. $\boxed{8}+9=17$
15. $4+8=12$ years
16. $9+2=11$ dolls
17. $6+\boxed{8}=14$ leaves
18. $7+2=9$ things
    $9+1=10$ things

Systematic Review 10F

1. $8+8=16$
2. $3+8=11$
3. $8+6=14$
4. $9+8=17$
5. $9+1=10$
6. $9+3=12$
7. $9+9=18$
8. $100+200=300$
9. $8+7=15$
10. $7+2=9$
11. $9+6=15$
12. $5+9=14$
13. $\boxed{9}+2=11$
14. $\boxed{8}+5=13$
15. $9+\boxed{7}=16$
16. $6+2=8$ years old
17. $8+8=16$ goodies
18. $5+2=7$ dollars
    $7+8=15$ dollars

Lesson Practice 11A

1. 6
2. 4
3. 2 circles
   3 triangles
4. 1 circle
   1 triangle
5. 2, 4, 6, 8, 10, 12
6. 2, 4, 6, 8, 10
7. circle
8. triangles

Lesson Practice 11B

1. 5
2. 6
3. 3
4. 7
5. 2, 4, 6, 8, 10, 12, 14
6. 2, 4, 6, 8, 10, 12, 14, 16
7. a triangle
8. a circle

Lesson Practice 11C

1. 4
2. 6
3. 5
4. 3
5. 7
6. 9
7. 3 circles
8. 4 triangles
9. 2, 4, 6, 8, 10, 12, 14, 16, 18, 20

Systematic Review 11D

1. 4 circles
   4 triangles
2. $8+3=11$
3. $5+8=13$
4. $8+9=17$
5. $4+8=12$
6. $9+7=16$
7. $9+6=15$
8. $2+3=5$
9. $10+70=80$
10. $4+2=6$
11. $7+8=15$
12. $9+9=18$
13. $6+0=6$
14. $8+\boxed{7}=15$
15. $\boxed{5}+2=7$
16. 2, 4, 6, 8, 10, 12, 14, 16, 18, 20

17. circles
18. 5 + 2 = 7 miles
 7 + [2] = 9 miles

Systematic Review 11E

1. 5 circles
 6 triangles
2. 8 + 5 = 13
3. 9 + 4 = 13
4. 6 + 1 = 7
5. 7 + 8 = 15
6. 8 + 8 = 16
7. 9 + 5 = 14
8. 2 + 8 = 10
9. 70 + 20 = 90
10. 8 + 0 = 8
11. 6 + 2 = 8
12. 8 + 4 = 12
13. 9 + 8 = 17
14. [8] + 6 = 14
15. 8 + [2] = 10
16. 2, 4, 6, 8, 10, 12, 14, 16, 18, 20
17. a triangle
18. 5 + 1 = 6 sweaters
 6 + [2] = 8 sweaters

Systematic Review 11F

1. 3 circles
 3 triangles
2. 3 + 8 = 11
3. 9 + 8 = 17
4. 9 + 9 = 18
5. 8 + 2 = 10
6. 300 + 100 = 400
7. 7 + 9 = 16
8. 4 + 0 = 4
9. 8 + 7 = 15
10. 9 + 3 = 12
11. 2 + 9 = 11
12. 6 + 8 = 14
13. 8 + 8 = 16
14. 9 + [7] = 16
15. 2 + [3] = 5
16. 2, 4, 6, 8, 10, 12, 14, 16, 18, 20
17. 9 sides
18. 6 + 1 = 7 cars
 7 + 9 = 16 cars

Lesson Practice 12A

1. 3 + 3 = 10 + 4 = 14
2. 4 + 4 = 3 + 3 = 6
3. 7 + 7 = 4 + 4 = 8
4. Five plus five equals ten.
5. Four plus four equals eight.
6. 2 + 2 = 4
7. 8 + 8 = 16
8. 3 + 3 = 6
9. 1 + 1 = 2
10. 6 + 6 = 12
11. 9 + 9 = 18
12. 7 + 7 = 14 children

Lesson Practice 12B

1. 5 + 5 = 10 —— 5 + 5 = 10
2. 8 + 8 = 16 10 + 2 = 12
3. 6 + 6 = 12 10 + 6 = 16
4. One plus one equals two.
5. Two plus two equals four.
6. 1 + 1 = 2
7. 9 + 9 = 18
8. 4 + 4 = 8
9. 3 + 3 = 6
10. 2 + 2 = 4
11. 5 + 5 = 10
12. 4 + 4 = 8 years

Lesson Practice 12C

1. 7 + 7 = 10 + 2 = 12
2. 6 + 6 = 4 + 4 = 8
3. 4 + 4 = 10 + 4 = 14
4. Three plus three equals six.
5. Five plus five equals ten.
6. 8 + 8 = 16

7. $2+2=4$
8. $7+7=14$
9. $4+4=8$
10. $1+1=2$
11. $9+9=18$
12. $6+6=12$ eggs

Systematic Review 12D

1. $40+40=80$
2. $7+7=14$
3. $3+3=6$
4. $6+6=12$
5. $8+3=11$
6. $4+8=12$
7. $9+7=16$
8. $6+8=14$
9. $5+5=10$
10. $4+2=6$
11. $9+9=18$
12. $8+1=9$
13. $\boxed{8}+8=16$
14. $\boxed{5}+9=14$
15. $\boxed{2}+5=7$
16. 4
17. 3
18. 2, 4, 6, 8, 10, 12, 14, 16, 18, 20
19. $8+\boxed{7}=15$ people
20. $3+2=5$ candles

Systematic Review 12E

1. $5+5=10$
2. $8+8=16$
3. $7+7=14$
4. $30+30=60$
5. $6+6=12$
6. $8+7=15$
7. $5+8=13$
8. $9+3=12$
9. $3+3=6$
10. $8+2=10$
11. $9+8=17$
12. $2+3=5$
13. $\boxed{2}+2=4$
14. $\boxed{9}+6=15$
15. $\boxed{1}+7=8$
16. 3 circles
17. 3 triangles
18. 2, 4, 6, 8, 10, 12, 14, 16, 18, 20
19. triangle
20. $9+\boxed{4}=13$ pennies

Systematic Review 12F

1. $8+8=16$
2. $6+6=12$
3. $200+200=400$
4. $5+5=10$
5. $4+4=8$
6. $7+7=14$
7. $8+6=14$
8. $4+8=12$
9. $8+9=17$
10. $9+5=14$
11. $6+2=8$
12. $7+0=7$
13. $\boxed{8}+3=11$
14. $\boxed{9}+9=18$
15. $\boxed{3}+3=6$
16. ○ ○ ○ ○
17. △ △ △
18. 10, 20, 30, 40, 50, 60, 70, 80, 90, 100
19. $8+8=16$ spider legs
20. $3+\boxed{2}=5$ inches

Lesson Practice 13A

1. 4
2. 7
3. 3
4. 3
5. 5, 10, 15, 20, 25, 30
6. rectangle
7. 4

Lesson Practice 13B

1. 7
2. 9
3. 4
4. 4
5. 5, 10, 15, 20, 25, 30, 35
6. a square
7. 4

Lesson Practice 13C

1. 6
2. 8
3. 3
4. 3
5. 4
6. 4
7. □ □
8. ▭ ▭ ▭
9. 5, 10, 15, 20, 25, 30, 35, 40, 45, 50

Systematic Review 13D

1. $6 + 6 = 12$
2. $8 + 8 = 16$
3. $7 + 9 = 16$
4. $8 + 3 = 11$
5. $4 + 4 = 8$
6. $9 + 8 = 17$
7. $50 + 20 = 70$
8. $7 + 7 = 14$
9. $1 + 7 = 8$
10. $5 + 5 = 10$
11. $8 + 6 = 14$
12. $9 + 5 = 14$
13. $\boxed{3} + 3 = 6$
14. $\boxed{8} + 4 = 12$
15. $\boxed{9} + 9 = 18$
16. 4
17. 2
18. 5, 10, 15, 20, 25, 30, 35, 40, 45, 50
19. 8 sides
20. $4 + 1 = 5$ pictures
    $5 + 5 = 10$ pictures

Systematic Review 13E

1. $3 + 3 = 6$
2. $50 + 10 = 60$
3. $9 + 4 = 13$
4. $8 + 7 = 15$
5. $6 + 6 = 12$
6. $4 + 0 = 4$
7. $4 + 4 = 8$
8. $3 + 9 = 12$
9. $7 + 7 = 14$
10. $8 + 5 = 13$
11. $6 + 2 = 8$
12. $9 + 9 = 18$
13. $\boxed{8} + 0 = 8$
14. $\boxed{6} + 6 = 12$
15. $\boxed{9} + 7 = 16$
16. 3 circles
17. 2 rectangles
18. 5, 10, 15, 20, 25, 30, 35, 40, 45, 50
19. $\boxed{3} + 9 = 12$ games
20. $6 + 2 = 8$ pies
    $8 + \boxed{1} = 9$ pies

Systematic Review 13F

1. $9 + 7 = 16$
2. $5 + 5 = 10$
3. $8 + 9 = 17$
4. $7 + 7 = 14$
5. $2 + 4 = 6$
6. $300 + 100 = 400$
7. $9 + 5 = 14$
8. $3 + 8 = 11$
9. $4 + 4 = 8$
10. $8 + 7 = 15$
11. $2 + 6 = 8$
12. $8 + 8 = 16$
13. $\boxed{2} + 1 = 3$
14. $\boxed{2} + 7 = 9$
15. $\boxed{1} + 5 = 6$

16. 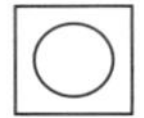

17.

18. 5, 10, 15, 20, 25, 30, 35, 40, 45, 50

19. [7] + 8 = 15 dollars

20. 6 + 1 = 7 pies
    7 + 7 = 14 pies

Lesson Practice 14A

1. 3 + 4 = 7
   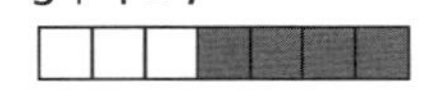
2. 4 + 5 = 9
   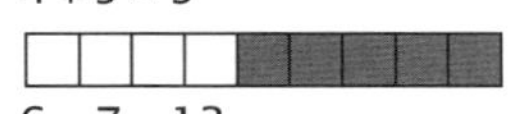
3. 6 + 7 = 13
   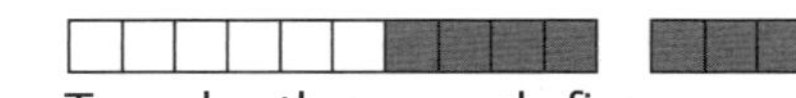
4. Two plus three equals five.
5. One plus two equals three.
6. 7 + 8 = 15
7. 1 + 2 = 3
8. 8 + 9 = 17
9. 4 + 5 = 9
10. 2 + 3 = 5
11. 7 + 8 = 15
12. 5 + 6 = 11 pets

Lesson Practice 14B

1. 1 + 2 = 3
   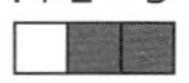
2. 7 + 8 = 15
   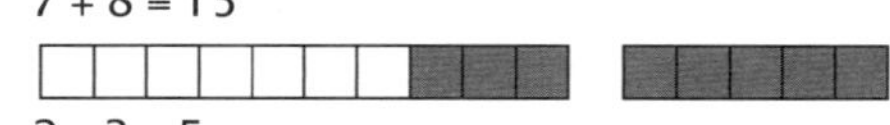
3. 2 + 3 = 5
   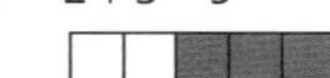
4. Four plus five equals nine.
5. Three plus four equals seven.
6. 5 + 6 = 11
7. 8 + 9 = 17
8. 6 + 7 = 13
9. 7 + 8 = 15
10. 4 + 3 = 7
11. 9 + 10 = 19
12. 6 + 7 = 13 children

Lesson Practice 14C

1. 5 + 6 = 11
   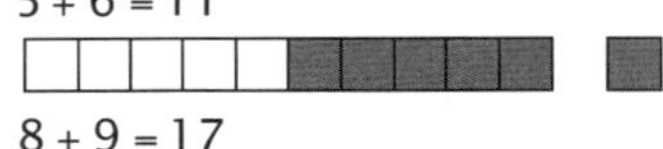
2. 8 + 9 = 17
   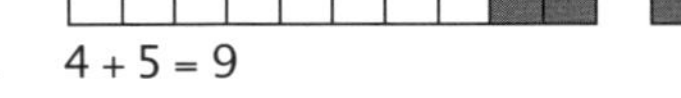
3. 4 + 5 = 9
4. One plus two equals three.
5. Two plus three equals five.
6. 2 + 3 = 5
7. 10 + 9 = 19
8. 6 + 7 = 13
9. 3 + 4 = 7
10. 9 + 8 = 17
11. 6 + 5 = 11
12. 7 + 8 = 15 leaves

Systematic Review 14D

1. 40 + 50 = 90
2. 6 + 7 = 13
3. 8 + 9 = 17
4. 7 + 8 = 15
5. 4 + 4 = 8
6. 8 + 8 = 16
7. 8 + 6 = 14
8. 9 + 5 = 14
9. 5 + 2 = 7
10. 7 + 1 = 8
11. 9 + 9 = 18
12. 2 + 3 = 5
13. 8 + 3 = 11
14. 9 + 7 = 16
15. Five plus five equals ten.
16. Zero plus zero equals zero.
17. square
18. triangle
19. 6 + 7 = 13 squirrels
20. 5, 10, 15, 20, 25, 30, 35, 40, 45, 50

Systematic Review 14E

1. $30+40=70$
2. $8+7=15$
3. $4+5=9$
4. $8+9=17$
5. $7+7=14$
6. $8+4=12$
7. $9+3=12$
8. $200+700=900$
9. $5+1=6$
10. $6+7=13$
11. $3+3=6$
12. $8+5=13$
13. $9+6=15$
14. $10+2=12$
15. Three plus two equals five.
16. Four plus four equals eight.
17. rectangle
18. circle
19. $3+4=7$ fingers
20. 2, 4, 6, 8, 10, 12, 14, 16, 18, 20

Systematic Review 14F

1. $5+6=11$
2. $50+40=90$
3. $6+7=13$
4. $9+8=17$
5. $2+2=4$
6. $8+9=17$
7. $9+4=13$
8. $600+200=800$
9. $10+8=18$
10. $7+1=8$
11. $9+9=18$
12. $2+3=5$
13. $9+7=16$
14. $4+2=6$
15. Six plus one equals seven.
16. One plus two equals three.
17. rectangle
18. square
19. $8+9=17$ sit-ups
20. 10, 20, 30, 40, 50, 60, 70, 80, 90, 100

Lesson Practice 15A

1. $1+9=10$

2. $6+4=10$
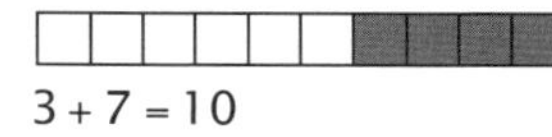
3. $3+7=10$
4. Five plus five equals ten.
5. Eight plus two equals ten.
6. $4+6=10$
7. $3+7=10$
8. $5+5=10$
9. $9+\boxed{1}=10$
10. $2+\boxed{8}=10$
11. $7+\boxed{3}=10$
12. $6+\boxed{4}=10$ apples

Lesson Practice 15B

1. $2+8=10$
2. $5+5=10$

3. $9+1=10$

4. Six plus four equals ten.
5. Seven plus three equals ten.
6. $5+5=10$
7. $9+1=10$
8. $6+4=10$
9. $3+\boxed{7}=10$
10. $6+\boxed{4}=10$
11. $5+\boxed{5}=10$
12. $8+\boxed{2}=10$ people

Lesson Practice 15C

1. $7+3=10$
2. $4+6=10$
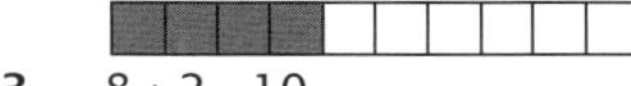
3. $8+2=10$

4. Three plus [seven] equals ten.
5. Five plus [five] equals ten.
6. 2 + 8 = 10
7. 7 + 3 = 10
8. 4 + 6 = 10
9. 1 + [9] = 10
10. 4 + [6] = 10
11. 8 + [2] = 10
12. 3 + [7] = 10 runs

Systematic Review 15D

1. 5 + 5 = 10
2. 3 + 7 = 10
3. 6 + 4 = 10
4. 200 + 300 = 500
5. 9 + 9 = 18
6. 8 + 6 = 14
7. 4 + 8 = 12
8. 9 + 3 = 12
9. 7 + 2 = 9
10. 6 + 7 = 13
11. 9 + 8 = 17
12. 2 + [8] = 10
13. 9 + [1] = 10
14. 3 + [4] = 7
15. 4 + [4] = 8
16. 8 + [8] = 16
17. 3 + [7] = 10
18. circles
19. 5 + [5] = 10 years
20. 9 + 4 = 13 models

Systematic Review 15E

1. 8 + 2 = 10
2. 9 + 1 = 10
3. 7 + 8 = 15
4. 4 + 6 = 10
5. 8 + 9 = 17
6. 20 + 50 = 70
7. 9 + 6 = 15
8. 8 + 3 = 11
9. 2 + 2 = 4
10. 5 + 6 = 11
11. 9 + 7 = 16
12. 7 + [3] = 10
13. 4 + [5] = 9
14. 1 + [9] = 10
15. 8 + [8] = 16
16. 6 + [4] = 10
17. 7 + [8] = 15
18. 2 + [8] = 10 years
19. 5 + 1 = 6 balls
    6 + [4] = 10 balls
20. 5, 10, 15, 20, 25, 30, 35, 40, 45, 50

Systematic Review 15F

1. 7 + 3 = 10
2. 60 + 20 = 80
3. 1 + 9 = 10
4. 8 + 5 = 13
5. 7 + 7 = 14
6. 5 + 9 = 14
7. 9 + 4 = 13
8. 6 + 6 = 12
9. 7 + 8 = 15
10. 3 + 4 = 7
11. 7 + 1 = 8
12. 6 + [0] = 6
13. 3 + [7] = 10
14. 8 + [2] = 10
15. 9 + [5] = 14
16. 8 + [3] = 11
17. 5 + [2] = 7
18. 2 + 8 = 10 times
19. 7 + [3] = 10 years
20. 6 + 2 = 8 chickens
    8 + 8 = 16 chickens

Lesson Practice 16A

1. 1 + 8 = 9
2. 6 + 3 = 9
3. 7 + 2 = 9

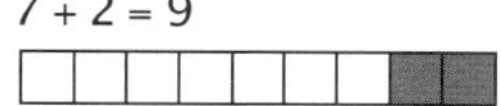

4. Nine plus zero equals nine.
5. Five plus four equals nine.
6. 8 + 1 = 9
7. 3 + 6 = 9
8. 9 + 0 = 9
9. 2 + 7 = 9
10. 4 + 5 = 9
11. 3 + 6 = 9
12. 7 + [2] = 9 books

Lesson Practice 16B

1. 5 + 4 = 9

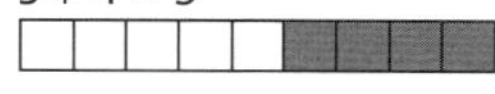

2. 8 + 1 = 9

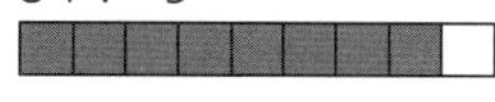

3. 2 + 7 = 9
4. Three plus six equals nine.
5. Eight plus one equals nine.
6. 6 + 3 = 9
7. 2 + 7 = 9
8. 1 + 8 = 9
9. 0 + 9 = 9
10. 5 + 4 = 9
11. 7 + 2 = 9
12. 6 + [3] = 9 cars

Lesson Practice 16C

1. 3 + 6 = 9
2. 4 + 5 = 9
3. 9 + 0 = 9

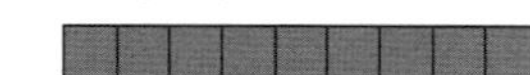

4. One plus eight equals nine.
5. Four plus five equals nine.
6. 7 + 2 = 9
7. 8 + 1 = 9
8. 3 + 6 = 9
9. 5 + 4 = 9
10. 2 + 7 = 9
11. 9 + 0 = 9
12. 5 + [4] = 9 candles

Systematic Review 16D

1. 8 + 1 = 9
2. 5 + 4 = 9
3. 60 + 30 = 90
4. 7 + 2 = 9
5. 9 + 9 = 18
6. 7 + 9 = 16
7. 8 + 4 = 12
8. 8 + 2 = 10
9. 3 + 2 = 5
10. 9 + 6 = 15
11. 6 + 7 = 13
12. 7 + [2] = 9
13. 8 + [6] = 14
14. 7 + [7] = 14

15\.
16\.

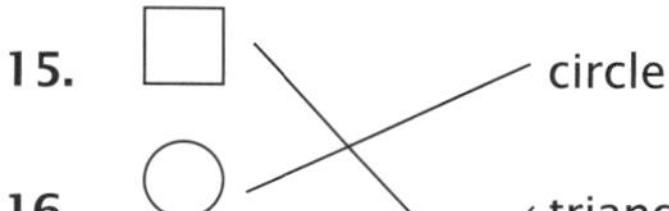

17\.
18\.

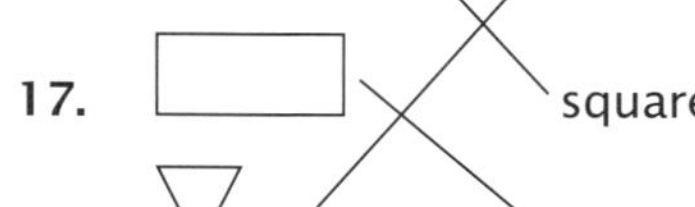

circle
triangle
square
rectangle

19. 5, 10, 15, 20, 25, 30, 35, 40, 45, 50
20. 3 + 1 = 4 rooms
    4 + 2 = 6 rooms

Systematic Review 16E

1. 6 + 6 = 12
2. 1 + 9 = 10
3. 50 + 40 = 90
4. 8 + 3 = 11
5. 9 + 6 = 15
6. 5 + 5 = 10
7. 5 + 6 = 11
8. 8 + 7 = 15
9. 5 + 2 = 7
10. 9 + 0 = 9
11. 3 + 1 = 4
12. $7+\boxed{3}=10$
13. $2+\boxed{7}=9$
14. $9+\boxed{9}=18$
15. 3 + 3 = 6 circles
16. 4 + 1 = 5 squares
17. 2 + 2 = 4 triangles
18. 5 + 0 = 5 rectangles
19. $4+\boxed{5}=9$ players
20. 3 + 4 = 7 flowers
    $7+\boxed{2}=9$ flowers

Systematic Review 16F

1. 7 + 3 = 10
2. 40 + 40 = 80
3. 9 + 1 = 10
4. 6 + 3 = 9
5. 9 + 5 = 14
6. 8 + 5 = 13
7. 3 + 2 = 5
8. 500 + 400 = 900
9. 5 + 6 = 11
10. 8 + 7 = 15
11. 9 + 8 = 17
12. $6+\boxed{4}=10$
13. $5+\boxed{8}=13$
14. $6+\boxed{6}=12$
15. 4 + 2 = 6
16. 5 + 4 = 9
17. 3 + 1 = 4
18. 2 + 6 = 8
19. 8 + 0 = 8 fireflies
20. 6 + 1 = 7 stories
    7 + 3 = 10 stories

Lesson Practice 17A

1. 4 + 7 = 11
2. 5 + 7 = 12
   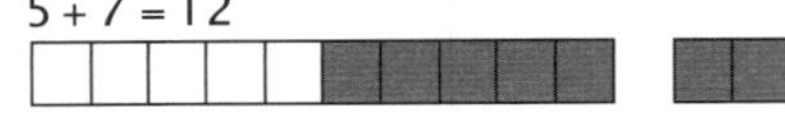
3. 3 + 5 = 8
   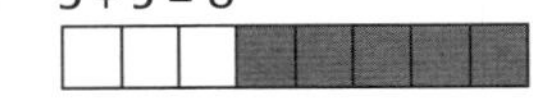
4. Five plus three equals eight.
5. Seven plus four equals eleven.
6. 4 + 7 = 11
7. 5 + 7 = 12
8. 3 + 5 = 8
9. 7 + 5 = 12
10. 5 + 3 = 8
11. 7 + 4 = 11
12. 5 + 7 = 12 students

Lesson Practice 17B

1. 5 + 3 = 8
2. 7 + 5 = 12
   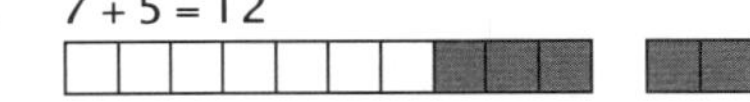
3. 7 + 4 = 11
   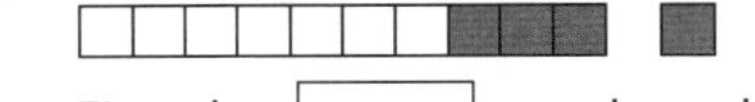
4. Five plus [seven] equals twelve.
5. Three plus five equals eight.
6. 5 + 3 = 8
7. 7 + 5 = 12
8. 7 + 4 = 11
9. 3 + 5 = 8
10. 4 + 7 = 11
11. 5 + 7 = 12
12. 3 + 5 = 8 quarts

Lesson Practice 17C

1. 7 + 4 = 11
2. 3 + 5 = 8
3. 5 + 7 = 12
4. Four plus [seven] equals eleven.
5. Seven plus [five] equals twelve.
6. 3 + 5 = 8
7. 4 + 7 = 11
8. 7 + 5 = 12
9. 7 + 4 = 11
10. 5 + 7 = 12
11. 5 + 3 = 8
12. 4 + 7 = 11 pictures

Systematic Review 17D

1. 5 + 7 = 12
2. 50 + 30 = 80
3. 8 + 6 = 14
4. 7 + 4 = 11
5. 7 + 3 = 10
6. 4 + 5 = 9
7. 4 + 7 = 11
8. 300 + 500 = 800
9. 8 + 2 = 10
10. 7 + 5 = 12
11. 6 + 4 = 10
12. 5 + [3] = 8
13. 3 + [4] = 7
14. 7 + [4] = 11
15. 6 + [5] = 11
16. 6 + [3] = 9
17. 4 + [4] = 8
18. 2, 4, 6, 8, 10, 12, 14, 16, 18, 20
19. 5 + 7 = 12 creatures
20. 4 + 2 + 4 = 10

Systematic Review 17E

1. 60 + 30 = 90
2. 7 + 4 = 11
3. 9 + 3 = 12
4. 6 + 4 = 10
5. 7 + 7 = 14
6. 7 + 5 = 12
7. 400 + 200 = 600
8. 4 + 7 = 11
9. 5 + 3 = 8
10. 9 + 1 = 10
11. 2 + 7 = 9
12. 8 + [9] = 17
13. 1 + [7] = 8
14. 8 + [2] = 10
15. 9 + [7] = 16
16. 5 + [6] = 11
17. 5 + [4] = 9
18. 5, 10, 15, 20, 25, 30, 35, 40, 45, 50
19. 3 + 5 = 8 pounds
20. 3 + 1 + 7 = 11

Systematic Review 17F

1. 7 + 3 = 10
2. 300 + 600 = 900
3. 8 + 4 = 12
4. 4 + 7 = 11
5. 5 + 0 = 5
6. 4 + 6 = 10
7. 5 + 7 = 12
8. 9 + 9 = 18
9. 5 + 4 = 9
10. 4 + 0 = 4
11. 4 + 3 = 7
12. 9 + [9] = 18
13. 5 + [7] = 12
14. 5 + [5] = 10
15. 3 + [7] = 10
16. 3 + [5] = 8
17. 7 + [6] = 13

18. 10, 20, 30, 40, 50, 60, 70, 80, 90, 100
19. 1 + 2 + 6 = 9
20. 4 + 7 = 11 shapes
11. 4 – 3 = 1
12. 5 – 1 = 4
13. 9 – 7 = 2
14. 6 – 4 = 2

Lesson Practice 18A

1. have 5 (blue)
2. owe 3 (pink)
3. have 7 (tan)
4. owe 6 (violet)
5. owe 1 (green)
6. have 2 (orange)
7. 8 – 3 = 5
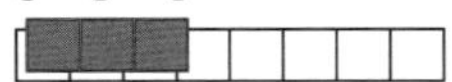
8. 7 – 4 = 3

9. 9 – 3 = 6
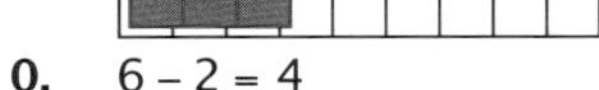
10. 6 – 2 = 4
11. done
12. 6 – 3 = 3
13. 8 – 2 = 6
14. 5 – 4 = 1

Lesson Practice 18B

1. have 3 (pink)
2. owe 5 (light blue)
3. owe 7 (tan)
4. have 4 (yellow)
5. owe 3 (pink)
6. have 1 (green)
7. 9 – 2 = 7
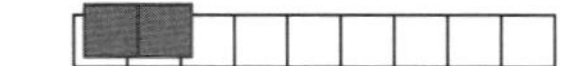
8. 6 – 5 = 1

9. 8 – 4 = 4
10. 3 – 1 = 2
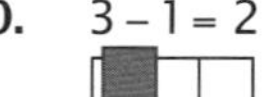

Lesson Practice 18C

1. have 8 (brown)
2. owe 9 (light green)
3. have 2 (orange)
4. owe 6 (violet)
5. owe 5 (light blue)
6. have 7 (tan)
7. 7 – 3 = 4
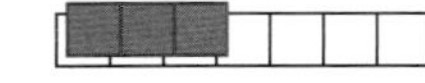
8. 5 – 4 = 1
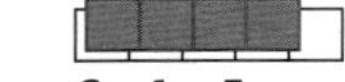
9. 6 – 1 = 5
10. 8 – 7 = 1
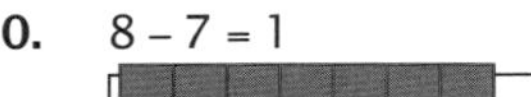
11. 9 – 5 = 4
12. 4 – 2 = 2
13. 7 – 6 = 1
14. 3 – 2 = 1

Systematic Review 18D

1. owe 3 (pink)
2. owe 5 (light blue)
3. have 6 (violet)
4. owe 2 (orange)
5. 2 – 1 = 1
6. 4 – 2 = 2
7. 5 + 3 = 8
8. 4 + 7 = 11
9. 30 + 40 = 70
10. 8 + 9 = 17
11. 6 + 6 = 12
12. 7 + 5 = 12
13. 6 + 3 = 9
14. 500 + 400 = 900
15. [8] + 0 = 8

16. [5] + 1 = 6
17. [9] + 0 = 9
18. [5] + 2 = 7
19. [2] + 8 = 10
20. 3 + 2 = 5 flowers

Systematic Review 18E

1. owe 4 (yellow)
2. have 8 (brown)
3. have 3 (pink)
4. owe 9 (light green)
5. 6 − 3 = 3
6. 7 − 5 = 2
7. 7 + 6 = 13
8. 40 + 40 = 80
9. 6 + 5 = 11
10. 300 + 500 = 800
11. 5 + 7 = 12
12. 7 + 4 = 11
13. 4 + 5 = 9
14. 5 + 6 = 11
15. [3] + 2 = 5
16. [3] + 1 = 4
17. [6] + 2 = 8
18. [4] + 0 = 4
19. [3] + 7 = 10
20. 5 + 6 = 11 books

Systematic Review 18F

1. have 7 (tan)
2. owe 6 (violet)
3. owe 10 (dark blue)
4. have 5 (light blue)
5. 8 − 5 = 3
6. 5 − 2 = 3
7. 4 + 7 = 11
8. 30 + 50 = 80
9. 5 + 7 = 12
10. 9 + 9 = 18
11. 7 + 5 = 12
12. 6 + 8 = 14
13. 7 + 4 = 11
14. 9 + 6 = 15
15. [6] + 1 = 7
16. [6] + 2 = 8
17. [8] + 1 = 9
18. [5] + 0 = 5
19. [4] + 5 = 9
20. 9 + 6 = 15 inches

Lesson Practice 19A

1. 10, 9, 8, 7, 6, 5, 4, 3, 2, 1, 0
2. done
3. 5 − 5 = 0
4. 1 − 0 = 1
5. 8 − 8 = 0
6. 6 − 5 = 1
7. 4 − 1 = 3
8. 9 − 1 = 8
9. 7 − 0 = 7
10. 4 − 3 = 1
11. 2 − 1 = 1
12. 10 − 1 = 9
13. 6 − 6 = 0
14. 9 − 8 = 1
15. 5 − 4 = 1
16. Eight minus one equals seven.
17. 2 − 0 = 2 kittens
18. 5 − 4 = 1 cookie

Lesson Practice 19B

1. 10, 9, 8, 7, 6, 5, 4, 3, 2, 1, 0
2. 5 − 4 = 1
3. 3 − 3 = 0
4. 0 − 0 = 0
5. 3 − 3 = 0
6. 6 − 1 = 5
7. 7 − 6 = 1
8. 10 − 9 = 1
9. 6 − 0 = 6
10. 3 − 1 = 2
11. 7 − 7 = 0

12. $4-0=4$
13. $8-7=1$
14. $7-1=6$
15. $1-1=0$
16. Nine minus zero equals nine.
17. $9-0=9$ pennies
18. $9-1=8$ dollars

Lesson Practice 19C

1. 10, 9, 8, 7, 6, 5, 4, 3, 2, 1, 0
2. $4-4=0$
3. $6-5=1$
4. $3-0=3$
5. $3-2=1$
6. $5-5=0$
7. $5-0=5$
8. $8-1=7$
9. $5-4=1$
10. $5-1=4$
11. $2-2=0$
12. $8-0=8$
13. $4-1=3$
14. $4-3=1$
15. $8-8=0$
16. Nine minus one equals eight.
17. $7-6=1$ year
18. $9-9=0$ pennies

Systematic Review 19D

1. $9-0=9$
2. $1-1=0$
3. $2-1=1$
4. $6-6=0$
5. $3-1=2$
6. $8-7=1$
7. $10-1=9$
8. $7-7=0$
9. $3+9=12$
10. $9+8=17$
11. $6+9=15$
12. $9+5=14$
13. $\boxed{7}+1=8$
14. $\boxed{4}+2=6$
15. $\boxed{4}+0=4$
16. $\boxed{3}+2=5$
17. Ten minus nine equals one.
18. Six minus zero equals six.
19. $6-1=5$ cards
20. $3-3=0$ toys

Systematic Review 19E

1. $3-3=0$
2. $7-1=6$
3. $7-6=1$
4. $4-0=4$
5. $1-1=0$
6. $6-1=5$
7. $9-8=1$
8. $7-0=7$
9. $9+1=10$
10. $9+7=16$
11. $2+9=11$
12. $4+9=13$
13. $\boxed{2}+2=4$
14. $\boxed{7}+2=9$
15. $\boxed{5}+2=7$
16. $\boxed{1}+2=3$
17. Four minus three equals one.
18. Five minus one equals four.
19. $6-5=1$ piece
20. $8-0=8$ friends

Systematic Review 19F

1. $5-4=1$
2. $4-1=3$
3. $8-8=0$
4. $5-0=5$
5. $3-2=1$
6. $8-1=7$
7. $4-4=0$
8. $1-0=1$
9. $9+0=9$
10. $9+9=18$
11. $5+9=14$

12. $8+9=17$
13. $\boxed{3}+2=5$
14. $\boxed{6}+2=8$
15. $\boxed{0}+2=2$
16. $\boxed{4}+2=6$
17. Eight minus seven equals one.
18. Three minus one equals two.
19. $7-7=0$ friends
20. $10-1=9$ years

Lesson Practice 20A

1. $5-2=3$
2. $5-3=2$
3. $7-2=5$
4. $7-5=2$
5. $10-2=8$
6. $10-8-2$
7. $30-20=10$
8. $300-100=200$
9. $8-2=6$
10. $8-6=2$
11. $4-2=2$
12. $2-2=0$
13. $11-2=9$
14. $11-9=2$
15. $7-5=2$ years
16. $9-2=7$ points

Lesson Practice 20B

1. $6-2=4$
2. $6-4=2$
3. $9-2=7$
4. $9-7=2$
5. $50-20=30$
6. $5-3=2$
7. $11-2=9$
8. $11-9=2$
9. $400-200=200$
10. $2-2=0$
11. $3-2=1$
12. $2-0=2$
13. $8-6=2$
14. $8-2=6$
15. $10-8=2$ stickers
16. $10-2=8$ dollars

Lesson Practice 20C

1. $4-2=2$
2. $3-1=2$
3. $10-8=2$
4. $10-2=8$
5. $7-5=2$
6. $70-20-50$
7. $6-2=4$
8. $6-4=2$
9. $8-6=2$
10. $800-200=600$
11. $5-2=3$
12. $5-3=2$
13. $2-2=0$
14. $3-2=1$
15. $9-7=2$ years
16. $6-2=4$ fish

Systematic Review 20D

1. $9-2=7$
2. $7-5=2$
3. $50-10=40$
4. $3-0=3$
5. $8-6=2$
6. $5-2=3$
7. $6-5=1$
8. $7-1=6$
9. $3+8=11$
10. $5+8=13$
11. $8+7=15$
12. $8+2=10$
13. $\boxed{9}+9=18$
14. $\boxed{6}+9=15$
15. $\boxed{2}+9=11$
16. $\boxed{4}+9=13$
17. $7-2=5$ pencils
18. $6-4=2$ people

19. 8 + 1 = 9 candy canes
20. 8 + 9 = 17 candy canes

Systematic Review 20E

1. 400 – 200 = 200
2. 9 – 7 = 2
3. 10 – 2 = 8
4. 11 – 9 = 2
5. 6 – 6 = 0
6. 10 – 9 = 1
7. 9 – 1 = 8
8. 2 – 0 = 2
9. 4 + 8 = 12
10. 8 + 6 = 14
11. 8 + 3 = 11
12. 7 + 8 = 15
13. $\boxed{3}$ + 9 = 12
14. $\boxed{5}$ + 9 = 14
15. $\boxed{8}$ + 9 = 17
16. $\boxed{1}$ + 9 = 10
17. 3 – 2 = 1 more boy
18. 10 – 8 = 2 people
19. 3 + 2 = 5 goats
20. 8 + 9 = 17 pages

Systematic Review 20F

1. 4 – 2 = 2
2. 7 – 2 = 5
3. 30 – 10 = 20
4. 5 – 3 = 2
5. 7 – 6 = 1
6. 2 – 2 = 0
7. 4 – 1 = 3
8. 9 – 8 = 1
9. 8 + 9 = 17
10. 8 + 0 = 8
11. 8 + 5 = 13
12. 8 + 8 = 16
13. $\boxed{7}$ + 9 = 16
14. $\boxed{9}$ + 9 = 18
15. $\boxed{4}$ + 9 = 13
16. $\boxed{6}$ + 9 = 15
17. 8 – 2 = 6 birds
18. 11 – 2 = 9 years
19. 7 + 4 = 11 days
20. 5 + 3 = 8 pounds

Lesson Practice 21A

1. 18 – 9 = 9
2. 9 – 9 = 0
3. 16 – 9 = 7
4. 15 – 9 = 6
5. 10 – 9 = 1
6. 12 – 9 = 3
7. 17 – 9 = 8
8. 11 – 9 = 2
9. 13 – 9 = 4
10. 18 – 9 = 9
11. 14 – 9 = 5
12. 9 – 9 = 0
13. 17 – 9 = 8 dollars
14. 12 – 9 = 3 cars
15. 13 – 9 = 4 eggs
16. 11 – 9 = 2 people

Lesson Practice 21B

1. 14 – 9 = 5
2. 10 – 9 = 1
3. 15 – 9 = 6
4. 11 – 9 = 2
5. 17 – 9 = 8
6. 13 – 9 = 4
7. 16 – 9 = 7
8. 9 – 9 = 0
9. 12 – 9 = 3
10. 10 – 9 = 1
11. 18 – 9 = 9
12. 14 – 9 = 5
13. 16 – 9 = 7 melons
14. 18 – 9 = 9 days
15. 15 – 9 = 6 dollars
16. 17 – 9 = 8 inches

Lesson Practice 21C

1. $9 - 9 = 0$
2. $14 - 9 = 5$
3. $11 - 9 = 2$
4. $13 - 9 = 4$
5. $16 - 9 = 7$
6. $10 - 9 = 1$
7. $12 - 9 = 3$
8. $15 - 9 = 6$
9. $18 - 9 = 9$
10. $9 - 9 = 0$
11. $17 - 9 = 8$
12. $11 - 9 = 2$
13. $10 - 9 = 1$ lollipop
14. $12 - 9 = 3$ eggs
15. $16 - 9 = 7$ dollars
16. $13 - 9 = 4$ years

Systematic Review 21D

1. $15 - 9 = 6$
2. $18 - 9 = 9$
3. $11 - 9 = 2$
4. $14 - 9 = 5$
5. $6 - 2 = 4$
6. $7 - 0 = 7$
7. $5 - 5 = 0$
8. $8 - 6 = 2$
9. $40 + 40 = 80$
10. $8 + 8 = 16$
11. $7 + 7 = 14$
12. $3 + 3 = 6$
13. $\boxed{5} + 8 = 13$
14. $\boxed{1} + 8 = 9$
15. $\boxed{7} + 8 = 15$
16. $\boxed{3} + 8 = 11$
17. $17 - 9 = 8$ minutes
18. $9 - 2 = 7$ granola bars
19. $5 + 7 = 12$ children
20. $5 + 3 = 8$ gifts
    $8 - 6 = 2$ gifts

Systematic Review 21E

1. $16 - 9 = 7$
2. $13 - 9 = 4$
3. $9 - 9 = 0$
4. $12 - 9 = 3$
5. $7 - 5 = 2$
6. $40 - 20 = 20$
7. $9 - 2 = 7$
8. $8 - 1 = 7$
9. $5 + 5 = 10$
10. $9 + 9 = 18$
11. $6 + 6 = 12$
12. $200 + 200 = 400$
13. $\boxed{2} + 8 = 10$
14. $\boxed{9} + 8 = 17$
15. $\boxed{4} + 8 = 12$
16. $\boxed{6} + 8 = 14$
17. $10 - 9 = 1$ bird
18. $14 - 9 = 5$ dollars
19. $4 + 5 = 9$ years
20. $6 + 4 = 10$ dollars
    $10 - 1 = 9$ dollars

Systematic Review 21F

1. $17 - 9 = 8$
2. $15 - 9 = 6$
3. $18 - 9 = 9$
4. $11 - 9 = 2$
5. $5 - 3 = 2$
6. $8 - 2 = 6$
7. $7 - 7 = 0$
8. $30 - 10 = 20$
9. $7 + 7 = 14$
10. $4 + 4 = 8$
11. $30 + 30 = 60$
12. $9 + 9 = 18$
13. $\boxed{8} + 8 = 16$
14. $\boxed{8} + 5 = 13$
15. $\boxed{8} + 0 = 8$
16. $\boxed{7} + 8 = 15$
17. $12 - 9 = 3$ questions
18. $8 - 6 = 2$ presents

19. 5 + 3 = 8 children
    8 − 2 = 6 children
20. 7 + 3 = 10 books

Lesson Practice 22A

1. 16 − 8 = 8
2. 12 − 8 = 4
3. 8 − 8 = 0
4. 10 − 8 = 2
5. 17 − 8 = 9
6. 13 − 8 = 5
7. 11 − 8 = 3
8. 15 − 8 = 7
9. 9 − 8 = 1
10. 14 − 8 = 6
11. 12 − 8 = 4
12. 17 − 8 = 9
13. 11 − 8 = 3 bananas
14. 13 − 8 = 5 books
15. 14 − 8 = 6 outfits
16. 16 − 8 = 8 dollars

Lesson Practice 22B

1. 10 − 8 = 2
2. 16 − 8 = 8
3. 9 − 8 = 1
4. 15 − 8 = 7
5. 8 − 8 = 0
6. 12 − 8 = 4
7. 14 − 8 = 6
8. 17 − 8 = 9
9. 11 − 8 = 3
10. 13 − 8 = 5
11. 10 − 8 = 2
12. 16 − 8 = 8
13. 9 − 8 = 1 puppy
14. 10 − 8 = 2 letters
15. 17 − 8 = 9 stickers
16. 12 − 8 = 4 gummy bears

Lesson Practice 22C

1. 13 − 8 = 5
2. 11 − 8 = 3
3. 15 − 8 = 7
4. 8 − 8 = 0
5. 16 − 8 = 8
6. 14 − 8 = 6
7. 12 − 8 = 4
8. 9 − 8 = 1
9. 17 − 8 = 9
10. 10 − 8 = 2
11. 16 − 8 = 8
12. 11 − 8 = 3
13. 8 − 8 = 0 snowmen
14. 15 − 8 = 7 days
15. 13 − 8 = 5 melons
16. 16 − 8 = 8 pounds

Systematic Review 22D

1. 14 − 8 = 6
2. 10 − 8 = 2
3. 13 − 8 = 5
4. 15 − 8 = 7
5. 12 − 9 = 3
6. 14 − 9 = 5
7. 70 − 20 = 50
8. 6 − 4 = 2
9. 6 + 4 = 10
10. 9 + 1 = 10
11. 8 + 2 = 10
12. 3 + 7 = 10
13. $\boxed{4}$ + 4 = 8
14. $\boxed{5}$ + 5 = 10
15. $\boxed{8}$ + 8 = 16
16. $\boxed{2}$ + 2 = 4
17. 17 − 8 = 9 grandsons
18. 12 − 8 = 4 eggs
19. 4 + 3 = 7 sandwiches
    7 − 1 = 6 sandwiches
20. 4 + 5 = 9 pencils
    9 − 2 = 7 pencils

Systematic Review 22E

1. $9-8=1$
2. $16-8=8$
3. $11-8=3$
4. $12-8=4$
5. $16-9=7$
6. $10-2=8$
7. $9-7=2$
8. $90-10=80$
9. $1+9=10$
10. $4+6=10$
11. $7+3=10$
12. $2+8=10$
13. $\boxed{6}+6=12$
14. $\boxed{3}+3=6$
15. $\boxed{7}+7=14$
16. $\boxed{9}+9=18$
17. $15-8=7$ minutes
18. $4-3=1$ books
19. $6+3=9$ runs
20. $10-8=2$ pennies
    $2+7=9$ pennies

Systematic Review 22F

1. $17-8=9$
2. $14-8=6$
3. $10-8=2$
4. $13-8=5$
5. $13-9=4$
6. $15-9=6$
7. $5-2=3$
8. $6-0=6$
9. $5+4=9$
10. $3+6=9$
11. $8+1=9$
12. $200+700=900$
13. $\boxed{1}+1=2$
14. $\boxed{4}+4=8$
15. $\boxed{8}+8=16$
16. $\boxed{5}+5=10$
17. $11-8=3$ marbles
18. $6+6=12$ chapters
19. $7+3=10$ years
20. $1+1=2$ lawns
    $2+3=5$ lawns

Lesson Practice 23A

1. $14-7=7$
2. $10-5=5$
3. $12-6=6$
4. $60-30=30$
5. $8-4=4$
6. $16-8=8$
7. $10-5=5$
8. $14-7=7$
9. $18-9=9$
10. $4-2=2$
11. $12-6=6$
12. $8-4=4$
13. $2-1=1$
14. $16-8=8$
15. $6-3=3$
16. $14-7=7$
17. $12-6=6$ years
18. $16-8=8$ apples

Lesson Practice 23B

1. $60-30=30$
2. $6-3=3$
3. $14-7=7$
4. $12-6=6$
5. $10-5=5$
6. $18-9=9$
7. $400-200=200$
8. $16-8=8$
9. $8-4=4$
10. $10-5=5$
11. $6-3=3$
12. $14-7=7$
13. $12-6=6$
14. $4-2=2$
15. $16-8=8$
16. $18-9=9$

17. 6 – 3 = 3 toys
18. 10 – 5 = 5 hours

Lesson Practice 23C

1. 10 – 5 = 5
2. 8 – 4 = 4
3. 16 – 8 = 8
4. 14 – 7 = 7
5. 6 – 3 = 3
6. 20 – 10 = 10
7. 18 – 9 = 9
8. 12 – 6 = 6
9. 14 – 7 = 7
10. 8 – 4 = 4
11. 10 – 5 = 5
12. 6 – 3 = 3
13. 18 – 9 = 9
14. 12 – 6 = 6
15. 4 – 2 = 2
16. 16 – 8 = 8
17. 14 – 7 = 7 miles
18. 4 – 2 = 2 more sisters

Systematic Review 23D

1. 12 – 6 = 6
2. 13 – 8 = 5
3. 8 – 4 = 4
4. 17 – 9 = 8
5. 14 – 7 = 7
6. 6 – 5 = 1
7. 10 – 5 = 5
8. 11 – 8 = 3
9. 6 – 3 = 3
10. 15 – 9 = 6
11. 9 – 1 = 8
12. 7 + 4 = 11
13. 500 + 300 = 800
14. 7 + 7 = 14
15. 2 + 8 = 10
16. $\boxed{6}$ + 4 = 10
17. $\boxed{7}$ + 3 = 10
18. $\boxed{1}$ + 9 = 10
19. 16 – 8 = 8 dollars
20. 7 + 2 = 9 turtles
    9 – 1 = 8 turtles

Systematic Review 23E

1. 6 – 3 = 3
2. 15 – 8 = 7
3. 10 – 5 = 5
4. 13 – 9 = 4
5. 12 – 6 = 6
6. 5 – 1 = 4
7. 80 – 40 = 40
8. 8 – 6 = 2
9. 14 – 7 = 7
10. 9 – 2 = 7
11. 5 – 3 = 2
12. 6 + 3 = 9
13. 50 + 40 = 90
14. 9 + 9 = 18
15. 7 + 2 = 9
16. $\boxed{2}$ + 8 = 10
17. $\boxed{4}$ + 6 = 10
18. $\boxed{3}$ + 7 = 10
19. 12 – 6 = 6 questions
20. 9 + 7 = 16 children

Systematic Review 23F

1. 18 – 9 = 9
2. 14 – 8 = 6
3. 12 – 6 = 6
4. 12 – 9 = 3
5. 8 – 4 = 4
6. 6 – 2 = 4
7. 600 – 300 = 300
8. 30 – 10 = 20
9. 10 – 5 = 5
10. 14 – 7 = 7
11. 16 – 9 = 7
12. 7 + 3 = 10
13. 1 + 9 = 10
14. 4 + 5 = 9

15. 8 + 3 = 11
16. [9] + 1 = 10
17. [5] + 5 = 10
18. [8] + 2 = 10
19. 9 − 9 = 0 plates
20. 8 + 3 = 11 people
    11 − 4 = 7 people

Lesson Practice 24A

1. 10 − 7 = 3
2. 10 − 4 = 6
3. 10 − 6 = 4
4. 10 − 3 = 7
5. 10 − 8 = 2
6. 10 − 4 = 6
7. 10 − 5 = 5
8. 10 − 2 = 8
9. 10 − 6 = 4
10. 10 − 0 = 10
11. 10 − 3 = 7
12. 10 − 9 = 1
13. 10 − 6 = 4
14. 10 − 1 = 9
15. 10 − 7 = 3
16. 10 − 3 = 7 stickers
17. 10 − 4 = 6 books
18. 10 − 7 = 3 lollipops

Lesson Practice 24B

1. 10 − 6 = 4
2. 10 − 2 = 8
3. 10 − 8 = 2
4. 10 − 5 = 5
5. 10 − 3 = 7
6. 10 − 7 = 3
7. 10 − 9 = 1
8. 10 − 4 = 6
9. 10 − 1 = 9
10. 10 − 5 = 5
11. 10 − 7 = 3
12. 10 − 6 = 4
13. 10 − 4 = 6
14. 10 − 0 = 10
15. 10 − 3 = 7
16. 10 − 6 = 4 dollars
17. 10 − 5 = 5 windows
18. 10 − 8 = 2 gallons

Lesson Practice 24C

1. 10 − 1 = 9
2. 10 − 3 = 7
3. 10 − 5 = 5
4. 10 − 9 = 1
5. 10 − 7 = 3
6. 10 − 6 = 4
7. 10 − 4 = 6
8. 10 − 8 = 2
9. 10 − 2 = 8
10. 10 − 3 = 7
11. 10 − 0 = 10
12. 10 − 5 = 5
13. 10 − 7 = 3
14. 10 − 6 = 4
15. 10 − 4 = 6
16. 10 − 4 = 6 years
17. 10 − 2 = 8 toads
18. 10 − 3 = 7 yards

Systematic Review 24D

1. 10 − 5 = 5
2. 12 − 6 = 6
3. 10 − 7 = 3
4. 17 − 8 = 9
5. 10 − 3 = 7
6. 70 − 20 = 50
7. 10 − 4 = 6
8. 14 − 7 = 7
9. 16 − 9 = 7
10. 10 − 8 = 2
11. 9 − 7 = 2
12. 8 + 5 = 13
13. 2 + 6 = 8
14. 3 + 4 = 7
15. 700 + 100 = 800

16. [4] + 5 = 9
17. [3] + 6 = 9
18. [8] + 1 = 9
19. 8 + 7 = 15 pieces
20. 5 + 4 = 9 calls
    9 + 9 = 18 calls

Systematic Review 24E

1. 10 – 7 = 3
2. 8 – 4 = 4
3. 10 – 6 = 4
4. 16 – 8 = 8
5. 10 – 5 = 5
6. 7 – 5 = 2
7. 10 – 3 = 7
8. 400 – 200 = 200
9. 12 – 8 = 4
10. 14 – 9 = 5
11. 10 – 1 = 9
12. 9 + 6 = 15
13. 6 + 5 = 11
14. 8 + 9 = 17
15. 30 + 50 = 80
16. [5] + 4 = 9
17. [7] + 2 = 9
18. [6] + 3 = 9
19. 10 – 8 = 2 students
20. 3 + 1 = 4 songs
    4 + 2 = 6 songs

Systematic Review 24F

1. 10 – 4 = 6
2. 6 – 3 = 3
3. 10 – 7 = 3
4. 9 – 8 = 1
5. 10 – 6 = 4
6. 11 – 9 = 2
7. 10 – 2 = 8
8. 600 – 400 = 200
9. 50 – 20 = 30
10. 4 – 4 = 0
11. 8 – 7 = 1
12. 2 + 9 = 11
13. 5 + 5 = 10
14. 7 + 6 = 13
15. 4 + 9 = 13
16. [1] + 8 = 9
17. [9] + 0 = 9
18. [2] + 7 = 9
19. 9 – 6 = 3 eggs
20. 5 + 2 = 7 bird houses
    9 – 7 = 2 bird houses

Lesson Practice 25A

1. 9 – 3 = 6
2. 9 – 6 = 3
3. 90 – 40 = 50
4. 9 – 7 = 2
5. 900 – 100 = 800
6. 9 – 9 = 0
7. 9 – 6 = 3
8. 9 – 5 = 4
9. 9 – 7 = 2
10. 9 – 4 = 5
11. 9 – 9 = 0
12. 9 – 3 = 6
13. 9 – 8 = 1
14. 9 – 2 = 7
15. 9 – 5 = 4
16. 9 – 6 = 3 dollars
17. 9 – 5 = 4 bandages
18. 9 – 4 = 5 gifts

Lesson Practice 25B

1. 9 – 8 = 1
2. 9 – 4 = 5
3. 9 – 5 = 4
4. 9 – 2 = 7
5. 9 – 6 = 3
6. 90 – 30 = 60
7. 900 – 700 = 200
8. 9 – 9 = 0
9. 9 – 1 = 8

10. 9 – 5 = 4
11. 9 – 8 = 1
12. 9 – 4 = 5
13. 9 – 2 = 7
14. 9 – 3 = 6
15. 9 – 6 = 3
16. 9 – 3 = 6 years
17. 9 – 1 = 8 sparrows
18. 9 – 5 = 4 chapters

Lesson Practice 25C

1. 9 – 7 = 2
2. 9 – 5 = 4
3. 9 – 3 = 6
4. 90 – 60 = 30
5. 9 – 0 = 9
6. 9 – 4 = 5
7. 900 – 200 = 700
8. 9 – 1 = 8
9. 9 – 8 = 1
10. 9 – 9 = 0
11. 9 – 5 = 4
12. 9 – 6 = 3
13. 9 – 3 = 6
14. 9 – 4 = 5
15. 9 – 7 = 2
16. 9 – 6 = 3 cows
17. 9 – 4 = 5 lines
18. 9 – 7 = 2 dollars

Systematic Review 25D

1. 9 – 6 = 3
2. 10 – 7 = 3
3. 9 – 4 = 5
4. 18 – 9 = 9
5. 90 – 50 = 40
6. 10 – 4 = 6
7. 9 – 3 = 6
8. 10 – 6 = 4
9. 8 – 2 = 6
10. 10 – 5 = 5
11. 6 – 4 = 2
12. 9 + 9 = 18
13. 200 + 300 = 500
14. 5 + 7 = 12
15. 6 + 6 = 12
16. $\boxed{3}$ + 4 = 7
17. $\boxed{5}$ + 3 = 8
18. $\boxed{4}$ + 3 = 7
19. 9 + 8 = 17 books
20. 6 + 7 = 13 flowers
    13 – 8 = 5 flowers

Systematic Review 25E

1. 9 – 2 = 7
2. 9 – 5 = 4
3. 10 – 3 = 7
4. 14 – 8 = 6
5. 9 – 3 = 6
6. 10 – 2 = 8
7. 13 – 8 = 5
8. 40 – 20 = 20
9. 12 – 8 = 4
10. 16 – 9 = 7
11. 15 – 8 = 7
12. 6 + 7 = 13
13. 50 + 40 = 90
14. 5 + 6 = 11
15. 7 + 7 = 14
16. $\boxed{5}$ + 3 = 8
17. $\boxed{4}$ + 3 = 7
18. $\boxed{3}$ + 5 = 8
19. 9 – 4 = 5 cards
20. 3 + 6 = 9 children
    11 – 9 = 2 children

Systematic Review 25F

1. 9 – 4 = 5
2. 10 – 8 = 2
3. 9 – 6 = 3
4. 16 – 8 = 8
5. 10 – 5 = 5
6. 13 – 9 = 4

7. $600 - 300 = 300$
8. $8 - 0 = 8$
9. $7 - 1 = 6$
10. $11 - 8 = 3$
11. $10 - 2 = 8$
12. $7 + 9 = 16$
13. $4 + 8 = 12$
14. $400 + 400 = 800$
15. $8 + 5 = 13$
16. $\boxed{4} + 3 = 7$
17. $\boxed{5} + 3 = 8$
18. $\boxed{3} + 4 = 7$
19. $9 + 7 = 16$ insects
20. $4 + 4 = 8$ pieces of fruit now
    $16 - 8 = 8$ pieces of fruit to buy

Lesson Practice 26A

1. $7 - 4 = 3$
2. $8 - 3 = 5$
3. $80 - 50 = 30$
4. $7 - 3 = 4$
5. $8 - 5 = 3$
6. $70 - 40 = 30$
7. $8 - 3 = 5$
8. $7 - 3 = 4$
9. $8 - 5 = 3$
10. $7 - 4 = 3$
11. $8 - 3 = 5$
12. $7 - 3 = 4$
13. Eight minus five equals three.
14. Seven minus four equals three.
15. $7 - 3 = 4$ monkeys
16. $8 - 5 = 3$ kittens
17. $7 - 4 = 3$ chores
18. $8 - 3 = 5$ inches

Lesson Practice 26B

1. $8 - 3 = 5$
2. $7 - 3 = 4$
3. $7 - 4 = 3$
4. $8 - 5 = 3$
5. $8 - 3 = 5$
6. $70 - 30 = 40$
7. $9 - 3 = 6$
8. $9 - 5 = 4$
9. $700 - 600 = 100$
10. $16 - 8 = 8$
11. $14 - 9 = 5$
12. $10 - 7 = 3$
13. Twelve minus six equals six.
14. Eleven minus nine equals two.
15. $7 - 3 = 4$ days
16. $7 - 4 = 3$ bags
17. $8 - 5 = 3$ feet
18. $8 - 3 = 5$ years

Lesson Practice 26C

1. $7 - 3 = 4$
2. $8 - 5 = 3$
3. $80 - 30 = 50$
4. $7 - 4 = 3$
5. $7 - 3 = 4$
6. $800 - 500 = 300$
7. $9 - 7 = 2$
8. $10 - 9 = 1$
9. $17 - 9 = 8$
10. $10 - 8 = 2$
11. $14 - 7 = 7$
12. $5 - 3 = 2$
13. Ten minus one equals nine.
14. Eight minus four equals four.
15. $8 - 3 = 5$ nickels
16. $7 - 3 = 4$ girls
17. $8 - 5 = 3$ chores
18. $8 - 3 = 5$ blocks

Systematic Review 26D

1. $8-5=3$
2. $7-4=3$
3. $8-3=5$
4. $7-3=4$
5. $9-4=5$
6. $70-20=50$
7. $5-0=5$
8. $12-9=3$
9. $4-4=0$
10. $10-4=6$
11. $14-8=6$
12. $7+8=15$
13. $500+400=900$
14. $3+2=5$
15. $0+9=9$
16. $\boxed{7}+4=11$
17. $\boxed{8}+7=15$
18. $\boxed{7}+5=12$
19. $8+2=10$ feet
20. $9+4=13$ dollars

Systematic Review 26E

1. $7-4=3$
2. $8-3=5$
3. $7-3=4$
4. $8-5=3$
5. $10-5=5$
6. $40-10=30$
7. $500-200=300$
8. $12-8=4$
9. $9-0=9$
10. $16-9=7$
11. $9-8=1$
12. $60+20=80$
13. $7+4=11$
14. $5+6=11$
15. $9+3=12$
16. $\boxed{7}+9=16$
17. $\boxed{6}+7=13$
18. $\boxed{4}+7=11$
19. $9+5=14$ players
20. $8+1=9$ steaks
    $9-2=7$ steaks

Systematic Review 26F

1. $8-3=5$
2. $8-7=1$
3. $7-3=4$
4. $80-40=40$
5. $9-6=3$
6. $11-8=3$
7. $7-1=6$
8. $10-6=4$
9. $6-4=2$
10. $7-5=2$
11. $15-9=6$
12. $9+6=15$
13. $3+8=11$
14. $400+200=600$
15. $9+8=17$
16. $\boxed{7}+8=15$
17. $\boxed{5}+7=12$
18. $\boxed{7}+6=13$
19. $5+5=10$ calls
20. $6+1=7$ bagels
    $7+6=13$ bagels

Lesson Practice 27A

1. $11-7=4$
2. $13-7=6$
3. $12-7=5$
4. $16-7=9$
5. $15-7=8$
6. $11-7=4$
7. $12-7=5$
8. $16-7=9$
9. $13-7=6$
10. $15-7=8$
11. $12-7=5$
12. $11-7=4$
13. Sixteen minus seven equals nine.
14. Thirteen minus seven equals six.

15. 13 − 7 = 6 doughnuts
16. 16 − 7 = 9 bulbs
17. 11 − 7 = 4 dollars
18. 15 − 7 = 8 miles

Lesson Practice 27B

1. 16 − 7 = 9
2. 11 − 7 = 4
3. 13 − 7 = 6
4. 15 − 7 = 8
5. 12 − 7 = 5
6. 16 − 7 = 9
7. 13 − 7 = 6
8. 10 − 3 = 7
9. 7 − 4 = 3
10. 8 − 2 = 6
11. 13 − 9 = 4
12. 8 − 5 = 3
13. Fifteen minus seven equals eight.
14. Eleven minus seven equals four.
15. 16 − 7 = 9 miles
16. 12 − 7 = 5 stories
17. 13 − 7 = 6 years
18. 15 − 7 = 8 cents

Lesson Practice 27C

1. 15 − 7 = 8
2. 12 − 7 = 5
3. 11 − 7 = 4
4. 13 − 7 = 6
5. 16 − 7 = 9
6. 12 − 7 = 5
7. 11 − 7 = 4
8. 14 − 7 = 7
9. 6 − 3 = 3
10. 13 − 8 = 5
11. 7 − 3 = 4
12. 18 − 9 = 9
13. Thirteen minus seven equals six.
14. Sixteen minus seven equals nine.
15. 15 − 7 = 8 pounds
16. 12 − 7 = 5 eggs
17. 16 − 7 = 9 people
18. 11 − 7 = 4 pounds

Systematic Review 27D

1. 12 − 7 = 5
2. 15 − 7 = 8
3. 11 − 7 = 4
4. 13 − 7 = 6
5. 8 − 6 = 2
6. 15 − 8 = 7
7. 80 − 30 = 50
8. 3 − 0 = 3
9. 17 − 8 = 9
10. 16 − 7 = 9
11. 10 − 2 = 8
12. 2 + 8 = 10
13. 3 + 5 = 8
14. 8 + 6 = 14
15. 400 + 300 = 700
16. $\boxed{6}$ + 5 = 11
17. $\boxed{7}$ + 6 = 13
18. $\boxed{9}$ + 6 = 15
19. 3 + 3 = 6 mice
20. 3 + 2 = 5 pets
    5 + 4 = 9 pets

Systematic Review 27E

1. 13 − 7 = 6
2. 11 − 7 = 4
3. 15 − 7 = 8
4. 12 − 7 = 5
5. 6 − 2 = 4
6. 14 − 8 = 6
7. 70 − 40 = 30
8. 10 − 6 = 4
9. 16 − 9 = 7
10. 11 − 2 = 9
11. 8 − 4 = 4
12. 6 + 4 = 10
13. 7 + 0 = 7

14. 9 + 8 = 17
15. 500 + 200 = 700
16. [6] + 8 = 14
17. [5] + 6 = 11
18. [6] + 7 = 13
19. 6 – 6 = 0 dollars
20. 18 – 9 = 9 years old

Systematic Review 27F

1. 11 – 7 = 4
2. 15 – 7 = 8
3. 13 – 7 = 6
4. 16 – 7 = 9
5. 12 – 7 = 5
6. 70 – 30 = 40
7. 10 – 4 = 6
8. 14 – 7 = 7
9. 11 – 4 = 7
10. 50 – 10 = 40
11. 17 – 8 = 9
12. 4 + 8 = 12
13. 9 + 2 = 11
14. 7 + 9 = 16
15. 8 + 3 = 11
16. [6] + 9 = 15
17. [8] + 6 = 14
18. [6] + 5 = 11
19. 3 + 3 = 6 good ideas
    6 + 7 = 13 good ideas
20. 3 + 6 = 9 gifts bought
    9 – 2 = 7 gifts left to wrap

Lesson Practice 28A

1. 14 – 6 = 8
2. 13 – 6 = 7
3. 11 – 6 = 5
4. 15 – 7 = 8
5. 12 – 6 = 6
6. 14 – 6 = 8
7. 13 – 6 = 7
8. 15 – 6 = 9
9. 10 – 6 = 4
10. 14 – 6 = 8
11. 11 – 6 = 5
12. 13 – 7 = 6
13. Fifteen minus six equals nine.
14. Eleven minus six equals five.
15. 13 – 6 = 7 children
16. 14 – 6 = 8 outfits
17. 15 – 6 = 9 daffodils
18. 11 – 6 = 5 peppers

Lesson Practice 28B

1. 11 – 6 = 5
2. 15 – 6 = 9
3. 14 – 6 = 8
4. 13 – 6 = 7
5. 11 – 6 = 5
6. 15 – 6 = 9
7. 9 – 4 = 5
8. 14 – 6 = 8
9. 13 – 6 = 7
10. 13 – 9 = 4
11. 8 – 3 = 5
12. 10 – 5 = 5
13. Eight minus five equals three.
14. Eleven minus eight equals three.
15. 14 – 6 = 8 days
16. 13 – 6 = 7 miles
17. 15 – 6 = 9 shells
18. 11 – 6 = 5 chores

Lesson Practice 28C

1. $13-6=7$
2. $12-6=6$
3. $15-6=9$
4. $11-6=5$
5. $14-6=8$
6. $13-6=7$
7. $7-5=2$
8. $6-3=3$
9. $15-7=8$
10. $12-8=4$
11. $16-7=9$
12. $6-1=5$
13. Fourteen minus six equals eight.
14. Nine minus six equals three.
15. $11-6=5$ people
16. $15-6=9$ balloons
17. $14-6=8$ dimes
18. $13-6=7$ apples

Systematic Review 28D

1. $15-6=9$
2. $11-6=5$
3. $14-6=8$
4. $13-6=7$
5. $10-3=7$
6. $16-7=9$
7. $15-9=6$
8. $10-7=3$
9. $90-50=40$
10. $14-9=5$
11. $15-8=7$
12. $40+30=70$
13. $6+4=10$
14. $8+5=13$
15. $7+7=14$
16. $\boxed{5}+7=12$
17. $\boxed{6}+5=11$
18. $\boxed{5}+9=14$
19. $8-1=7$ doughnuts
    $7-1=6$ doughnuts
20. $9+5=14$ people invited
    $14-7=7$ people didn't come

Systematic Review 28E

1. $14-6=8$
2. $11-6=5$
3. $13-6=7$
4. $15-6=9$
5. $13-7=6$
6. $17-9=8$
7. $50-30=20$
8. $8-6=2$
9. $12-7=5$
10. $15-7=8$
11. $12-9=3$
12. $3+7=10$
13. $3+9=12$
14. $8+6=14$
15. $30+50=80$
16. $\boxed{5}+6=11$
17. $\boxed{5}+8=13$
18. $\boxed{7}+5=12$
19. $7+4=11$ mosquitoes
    $11-9=2$ mosquitoes
20. $5+8=13$ things

Systematic Review 28F

1. $11-6=5$
2. $15-6=9$
3. $14-6=8$
4. $13-6=7$
5. $11-7=4$
6. $13-8=5$
7. $500-200=300$
8. $8-0=8$
9. $9-5=4$
10. $10-4=6$
11. $9-3=6$
12. $9+2=11$
13. $7+5=12$
14. $8+8=16$
15. $7+8=15$

16. $\boxed{8} + 5 = 13$
17. $\boxed{9} + 5 = 14$
18. $\boxed{6} + 5 = 11$
19. 7 + 7 = 14 feathers
    14 − 9 = 5 feathers
20. 7 + 4 = 11 glasses
    11 − 6 = 5 glasses

Lesson Practice 29A

1. 12 − 5 = 7
2. 11 − 5 = 6
3. 13 − 5 = 8
4. 14 − 5 = 9
5. 10 − 5 = 5
6. 11 − 5 = 6
7. 13 − 5 = 8
8. 12 − 5 = 7
9. 14 − 5 = 9
10. 12 − 5 = 7
11. 13 − 5 = 8
12. 11 − 5 = 6
13. Fourteen minus five equals nine.
14. Ten minus five equals five.
15. 14 − 5 = 9 dollars
16. 11 − 5 = 6 times
17. 13 − 5 = 8 dollars
18. 12 − 5 = 7 months

Lesson Practice 29B

1. 14 − 5 = 9
2. 12 − 5 = 7
3. 11 − 5 = 6
4. 13 − 5 = 8
5. 11 − 7 = 4
6. 14 − 6 = 8
7. 12 − 5 = 7
8. 14 − 5 = 9
9. 13 − 8 = 5
10. 10 − 2 = 8
11. 12 − 5 = 7
12. 6 − 3 = 3
13. Eleven minus five equals six.
14. Twelve minus seven equals five.
15. 13 − 5 = 8 feet
16. 14 − 5 = 9 years old
17. 12 − 5 = 7 models
18. 11 − 5 = 6 inches

Lesson Practice 29C

1. 13 − 5 = 8
2. 14 − 5 = 9
3. 12 − 5 = 7
4. 11 − 5 = 6
5. 8 − 4 = 4
6. 7 − 6 = 1
7. 8 − 3 = 5
8. 18 − 9 = 9
9. 15 − 8 = 7
10. 12 − 6 = 6
11. 9 − 3 = 6
12. 6 − 4 = 2
13. Sixteen minus nine equals seven.
14. Ten minus six equals four.
15. 14 − 5 = 9 people
16. 12 − 5 = 7 eggs
17. 11 − 5 = 6 miles
18. 13 − 5 = 8 dollars

Systematic Review 29D

1. 11 − 5 = 6
2. 13 − 5 = 8
3. 12 − 5 = 7
4. 14 − 5 = 9
5. 10 − 5 = 5
6. 13 − 9 = 4
7. 11 − 6 = 5
8. 9 − 4 = 5
9. 80 − 50 = 30
10. 16 − 7 = 9
11. 12 − 9 = 3
12. 6 + 7 = 13
13. 40 + 40 = 80
14. 8 + 6 = 14

15. 3 + 7 = 10
16. $\boxed{4}$ + 7 = 11
17. $\boxed{3}$ + 9 = 12
18. $\boxed{9}$ + 4 = 13
19. 7 − 5 = 2 flowers
    2 + 8 = 10 flowers
20. 7 + 5 = 12 hyenas
    12 − 9 = 3 hyenas

Systematic Review 29E

1. 12 − 5 = 7
2. 14 − 5 = 9
3. 11 − 5 = 6
4. 13 − 5 = 8
5. 15 − 6 = 9
6. 12 − 8 = 4
7. 60 − 20 = 40
8. 700 − 400 = 300
9. 10 − 7 = 3
10. 17 − 8 = 9
11. 9 − 6 = 3
12. 0 + 2 = 2
13. 9 + 7 = 16
14. 6 + 3 = 9
15. 7 + 8 = 15
16. $\boxed{4}$ + 8 = 12
17. $\boxed{8}$ + 3 = 11
18. $\boxed{7}$ + 4 = 11
19. 2 + 1 = 3 zoos
    3 + 5 = 8 zoos
20. 10 + 6 = 16 books
    16 − 8 = 8 books

Systematic Review 29F

1. 14 − 5 = 9
2. 11 − 5 = 6
3. 13 − 5 = 8
4. 12 − 5 = 7
5. 7 − 3 = 4
6. 17 − 9 = 8
7. 10 − 1 = 9
8. 13 − 7 = 6
9. 11 − 8 = 3
10. 7 − 5 = 2
11. 15 − 9 = 6
12. 9 + 8 = 17
13. 700 + 200 = 900
14. 8 + 0 = 8
15. 9 + 9 = 18
16. $\boxed{4}$ + 9 = 13
17. $\boxed{9}$ + 3 = 12
18. $\boxed{3}$ + 8 = 11
19. 6 + 1 = 7 hours
    7 − 4 = 3 hours
20. 8 + 6 = 14 times
    14 − 5 = 9 times

Lesson Practice 30A

1. 13 − 4 = 9
2. 12 − 4 = 8
3. 11 − 4 = 7
4. 11 − 3 = 8
5. 12 − 3 = 9
6. 12 − 4 = 8
7. 12 − 3 = 9
8. 13 − 4 = 9
9. 11 − 3 = 8
10. 13 − 4 = 9
11. 11 − 4 = 7
12. 12 − 3 = 9
13. Eleven minus three equals eight.
14. Twelve minus four equals eight.
15. 11 − 4 = 7 students
16. 13 − 4 = 9 days
17. 12 − 4 = 8 years
18. 12 − 3 = 9 lilies

Lesson Practice 30B

1. $11-3=8$
2. $13-4=9$
3. $12-3=9$
4. $12-4=8$
5. $11-4=7$
6. $13-4=9$
7. $13-5=8$
8. $12-4=8$
9. $12-3=9$
10. $12-5=7$
11. $11-4=7$
12. $11-3=8$
13. Fourteen minus nine equals five.
14. Eleven minus five equals six.
15. $12-4=8$ feet
16. $11-3=8$ apples
17. $13-4=9$ gallons
18. $11-4=7$ miles

Lesson Practice 30C

1. $12-4=8$
2. $11-3=9$
3. $11-4=7$
4. $12-3=9$
5. $13-4=9$
6. $14-5=9$
7. $10-8=2$
8. $11-9=2$
9. $14-8=6$
10. $11-4=7$
11. $11-3=8$
12. $12-4=8$
13. Thirteen minus four equals nine.
14. Twelve minus three equals nine.
15. $11-3=8$ girls
16. $12-4=8$ years
17. $12-3=9$ dimes
18. $13-4=9$ spaces

Systematic Review 30D

1. $12-3=9$
2. $11-4=7$
3. $13-4=9$
4. $11-3=8$
5. $12-4=8$
6. $16-8=8$
7. $10-3=7$
8. $9-7=2$
9. $3-3=0$
10. $8-4=4$
11. $11-5=6$
12. $13-6=7$
13. $9-5=4$
14. $15-6=9$
15. $2+3=5$
16. $9+1=10$
17. $60+30=90$
18. $200+700=900$
19. $11-4=7$ trees
20. $7+8=15$ pictures
    $15-9=6$ pictures

Systematic Review 30E

1. $11-4=7$
2. $13-4=9$
3. $11-3=8$
4. $12-4=8$
5. $12-3=9$
6. $14-7=7$
7. $6-3=3$
8. $13-5=8$
9. $14-6=8$
10. $10-7=3$
11. $12-8=4$
12. $14-5=9$
13. $90-40=50$
14. $8-5=3$
15. $4+7=11$
16. $8+3=11$
17. $9+4=13$
18. $6+5=11$

19. 15 − 9 = 6 math facts

20. 4 + 4 = 8 bouquets
8 − 3 = 5 bouquets

Systematic Review 30F

1. 13 − 4 = 9
2. 12 − 4 = 8
3. 12 − 3 = 9
4. 11 − 4 = 7
5. 11 − 3 = 8
6. 12 − 5 = 7
7. 11 − 6 = 5
8. 13 − 8 = 5
9. 10 − 4 = 6
10. 9 − 3 = 6
11. 12 − 7 = 5
12. 7 − 4 = 3
13. 17 − 9 = 8
14. 15 − 7 = 8
15. 8 + 4 = 12
16. 30 + 40 = 70
17. 7 + 9 = 16
18. 5 + 7 = 12
19. 5 + 6 = 11 gummy bears
11 − 7 = 4 gummy bears
20. 12 − 7 = 5 bananas
5 + 2 = 7 bananas

Appendix A1

1. 0:10
2. 0:20
3. 0:40
4. 0:35
5. 0:50
6. 0:45

Appendix A2

1. 0:05
2. 0:25
3. 0:55
4. 0:50
5. 0:30
6. 0:15

Appendix B1

1. 9:00
2. 7:00
3. 11:00
4. 1:00

Appendix B2

1. 8:00
2. 2:00
3. 12:00
4. 4:00

Appendix B3

1. 11:30
2. 5:45
3. 3:40
4. 12:25

Appendix B4

1. 5:50
2. 2:40
3. 7:10
4. 4:15

Test Solutions

Lesson Test 1

1. 5 hundreds, 3 tens, and 6 units;
   five hundred thirty-six
2. 129; one hundred twenty-nine
3. 1 hundred, 4 tens, and 1 unit;
   one hundred forty-one
4. 3 hundreds and 9 tens;
   three hundred ninety
5. 9
6. 9

Lesson Test 2

1. 0, 1, 2, 3, 4, 5, 6, 7, 8, 9, 10, 11,
   12, 13, 14, 15, 16, 17, 18, 19, 20
2. 0, 1, 2, 3, 4, 5, 6, 7, 8, 9, 10, 11,
   12, 13, 14, 15, 16, 17, 18, 19, 20
3. 263; two hundred sixty-three
4. 1 hundred, 5 tens, and 7 units;
   one hundred fifty-seven
5. 3 tens and 8 units;
   thirty-eight

Lesson Test 3

1. (5)
2. (6)
3. (7)
4. (4)
5. brown
6. 326; three hundred twenty-six
7. 0, 1, 2, 3, 4, 5, 6, 7, 8, 9, 10, 11,
   12, 13, 14, 15, 16, 17, 18, 19, 20

Lesson Test 4

1. $0+1=1$
2. $8+0=8$
3. $0+6=6$
4. $5+0=5$
5. $4+0=4$
6. $0+3=3$
7. $0+0=0$
8. $0+7=7$
9. $2+0=2$
10. $1+0=1$
11. $3+0=3$
12. $0+5=5$
13. 3 hundreds and 5 tens;
    three hundred fifty
14. 1 hundred and 2 units;
    one hundred two
15. light green
16. $6+0=6$

Lesson Test 5

1. $5+1=6$
2. $4+1=5$
3. $1+8=9$
4. $2+1=3$
5. $7+1=8$
6. $1+3=4$
7. $1+1=2$
8. $1+9=10$
9. $6+1=7$
10. $0+5=5$
11. $8+0=8$
12. $0+2=2$
13. 194; one hundred ninety-four
14. 2 hundreds, 1 ten, and 4 units;
    two hundred fourteen
15. $5+1=6$ children

Lesson Test 6

1.

0	1	2	3	4	5	6	7	8	9
10	11	12	13	14	15	16	17	18	19
20	21	22	23	24	25	26	27	28	29
30	31	32	33	34	35	36	37	38	39
40	41	42	43	44	45	46	47	48	49
50	51	52	53	54	55	56	57	58	59
60	61	62	63	64	65	66	67	68	69
70	71	72	73	74	75	76	77	78	79
80	81	82	83	84	85	86	87	88	89
90	91	92	93	94	95	96	97	98	99
100									

2. 10, 20, 30, 40, 50, 60, 70, 80, 90, 100
3. $1+6=7$
4. $5+1=6$
5. $9+0=9$
6. $1+8=9$

Lesson Test 7

1. $1+2=3$
2. $2+4=6$
3. $20+20=40$
4. $100+100=200$
5. $6+2=8$
6. $5+2=7$
7. $2+7=9$
8. $4+2=6$
9. $2+3=5$
10. $2+6=8$
11. $0+2=2$
12. $1+7=8$
13. $3+0=3$
14. $6+1=7$
15. $9+1=10$
16. $0+4=4$
17. $1+2=3$
18. $3+2=5$ pencils

Lesson Test 8

1. $\boxed{0}+1=1$
2. $\boxed{4}+0=4$
3. $\boxed{0}+2=2$
4. $\boxed{8}+0=8$
5. $\boxed{3}+2=5$
6. $\boxed{6}+1=7$
7. $\boxed{1}+2=3$
8. $\boxed{5}+1=6$
9. $\boxed{0}+0=0$
10. $\boxed{8}+1=9$
11. $\boxed{1}+0=1$
12. $\boxed{2}+2=4$
13. $6+2=8$
14. $7+1=8$
15. $30+20=50$
16. $9+0=9$
17. $\boxed{2}+3=5$ cars
18. $\boxed{1}+8=9$ players

Lesson Test 9

1. $0+9=9$
2. $9+7=16$
3. $6+9=15$
4. $9+9=18$
5. $9+2=11$
6. $3+9=12$
7. $9+1=10$
8. $8+9=17$
9. $9+7=16$
10. $4+9=13$
11. $6+1=7$
12. $7+2=9$
13. $\boxed{2}+9=11$
14. $\boxed{4}+2=6$
15. $\boxed{3}+1=4$
16. 10, 20, 30, 40, 50, 60, 70, 80, 90, 100
17. $9+8=17$ books
18. $6+\boxed{2}=8$ dollars

Lesson Test 10

1. $8+8=16$
2. $8+5=13$
3. $0+8=8$
4. $2+8=10$
5. $8+6=14$
6. $8+3=11$
7. $80+10=90$
8. $7+8=15$
9. $1+7=8$
10. $8+9=17$
11. $2+5=7$
12. $9+7=16$
13. $1+3=4$
14. $9+6=15$
15. $4+9=13$
16. $7+2=9$
17. $\boxed{8}+7=15$
18. $\boxed{9}+9=18$
19. $8+5=13$ miles
20. $\boxed{8}+9=17$ beads
21. $2+8=10$
22. $9+9=18$
23. $3+9=12$
24. $9+8=17$
25. $\boxed{8}+7=15$
26. $\boxed{9}+9=18$
27. $\boxed{9}+8=17$
28. 346; three hundred forty-six
29. 0, 1, 2, 3, 4, 5, 6, 7, 8, 9, 10, 11,
12, 13, 14, 15, 16, 17, 18, 19, 20
30. 10, 20, 30, 40, 50, 60, 70, 80, 90, 100
31. $8+3=11$ birds
32. $5+1=6$ pennies

Unit Test I

1. $7+9=16$
2. $2+2=4$
3. $4+9=13$
4. $2+5=7$
5. $6+2=8$
6. $8+7=15$
7. $20+10=30$
8. $8+0=8$
9. $2+9=11$
10. $9+9=18$
11. $70+20=90$
12. $100+800=900$
13. $5+8=13$
14. $8+4=12$
15. $8+8=16$
16. $2+4=6$
17. $8+6=14$
18. $6+9=15$
19. $2+3=5$
20. $9+5=14$

Lesson Test 11

1. 4
2. 4
3. 6
4. 4
5. $9+8=17$
6. $8+5=13$
7. $2+7=9$
8. $40+20=60$
9. $3+0=3$
10. $4+2=6$
11. $9+9=18$
12. $6+2=8$
13. $\boxed{6}+8=14$
14. $\boxed{5}+2=7$
15. 2, 4, 6, 8, 10, 12, 14, 16, 18, 20

Lesson Test 12

1. $8+8=16$
2. $5+5=10$
3. $30+30=60$
4. $4+4=8$
5. $6+6=12$
6. $9+9=18$
7. $7+7=14$
8. $2+2=4$
9. $5+8=13$

10. $9+4=13$
11. $3+2=5$
12. $6+9=15$
13. $\boxed{9}+9=18$
14. $\boxed{7}+8=15$
15. $\boxed{9}+2=11$
16. 3
17. 4
18. $4+4=8$ fish
19. $6+\boxed{9}=15$ questions
20. $7+7=14$ chores

Lesson Test 13

1. 3
2. 2
3. 3
4. 4
5. $6+6=12$
6. $3+3=6$
7. $7+7=14$
8. $20+40=60$
9. $9+0=9$
10. $7+1=8$
11. $8+9=17$
12. $5+8=13$
13. $\boxed{6}+9=15$
14. $4+\boxed{4}=8$
15. 4
16. 5, 10, 15, 20, 25, 30, 35, 40, 45, 50

Lesson Test 14

1. $50+40=90$
2. $8+7=15$
3. $1+2=3$
4. $6+7=13$
5. $2+3=5$
6. $5+6=11$
7. $8+9=17$
8. $400+300=700$
9. $7+7=14$
10. $8+3=11$
11. $6+6=12$
12. $9+5=14$
13. Three plus two equals five.
14. Two plus two equals four.
15. square
16. circle
17. $5+5=10$ toes
18. $5+1=6$ brothers and sisters
    $6+1=7$ children

Lesson Test 15

1. $6+4=10$
2. $9+7=16$
3. $10+20=30$
4. $5+5=10$
5. $9+1=10$
6. $6+8=14$
7. $8+2=10$
8. $8+8=16$
9. $7+3=10$
10. $3+4=7$
11. $5+2=7$
12. $6+4=10$
13. $5+6=11$
14. $7+8=15$
15. $9+5=14$
16. $7+6=13$
17. $\boxed{4}+6=10$
18. $\boxed{5}+5=10$
19. $\boxed{7}+3=10$
20. $8+\boxed{2}=10$ toys

Lesson Test 16

1. $2+7=9$
2. $5+5=10$
3. $10+80=90$
4. $6+3=9$
5. $7+8=15$
6. $5+4=9$
7. $6+6=12$
8. $5+9=14$
9. $8+1=9$

10. $4 + 4 = 8$
11. $8 + 4 = 12$
12. $5 + \boxed{4} = 9$
13. $6 + \boxed{3} = 9$
14. $7 + \boxed{2} = 9$
15. square
16. circle
17. rectangle
18. triangle
19. $4 + \boxed{5} = 9$ planets
20. $60 + 30 = 90$ pennies

Lesson Test 17

1. $30 + 50 = 80$
2. $4 + 7 = 11$
3. $7 + 5 = 12$
4. $5 + 4 = 9$
5. $5 + 7 = 12$
6. $7 + 3 = 10$
7. $200 + 200 = 400$
8. $8 + 7 = 15$
9. $5 + 5 = 10$
10. $9 + 9 = 18$
11. $6 + 4 = 10$
12. $3 + 8 = 11$
13. $6 + 3 = 9$
14. $8 + 5 = 13$
15. $4 + \boxed{7} = 11$
16. $7 + \boxed{5} = 12$
17. $3 + \boxed{5} = 8$
18. $6 + 5 = 11$ balloons
19. $4 + 3 = 7$ birds
20. 5, 10, 15, 20, 25, 30, 35, 40, 45, 50

Unit Test II

1. $9 + 2 = 11$
2. $3 + 3 = 6$
3. $7 + 6 = 13$
4. $3 + 5 = 8$
5. $5 + 1 = 6$
6. $2 + 7 = 9$
7. $40 + 10 = 50$
8. $2 + 4 = 6$
9. $2 + 2 = 4$
10. $2 + 8 = 10$
11. $50 + 20 = 70$
12. $400 + 500 = 900$
13. $8 + 6 = 14$
14. $9 + 7 = 16$
15. $40 + 30 = 70$
16. $300 + 100 = 400$
17. $8 + 3 = 11$
18. $6 + 9 = 15$
19. $6 + 6 = 12$
20. $8 + 4 = 12$
21. $0 + 6 = 6$
22. $7 + 7 = 14$
23. $5 + 6 = 11$
24. $8 + 5 = 13$
25. $7 + 3 = 10$
26. $9 + 9 = 18$
27. $9 + 4 = 13$
28. $7 + 8 = 15$
29. $4 + 4 = 8$
30. $2 + 3 = 5$
31. $20 + 30 = 50$
32. $7 + 5 = 12$
33. $0 + 9 = 9$
34. $5 + 5 = 10$
35. $10 + 10 = 20$
36. $600 + 300 = 900$
37. $8 + 0 = 8$
38. $3 + 9 = 12$
39. $20 + 60 = 80$
40. $300 + 300 = 600$
41. $8 + 8 = 16$
42. $4 + 7 = 11$
43. $9 + 8 = 17$
44. $4 + 6 = 10$
45. rectangle
46. triangle
47. square

Lesson Test 18

1. have 2 (orange)
2. owe 10 (dark blue)
3. have 3 (pink)
4. have 5 (light blue)
5. have 6 (violet)
6. owe 8 (brown)
7. $4+7=11$
8. $6+5=11$
9. $5+4=9$
10. $600+300=900$
11. $\boxed{7}+0=7$
12. $\boxed{2}+1=3$
13. $\boxed{6}+0=6$
14. $\boxed{8}+1=9$
15. $\boxed{5}+3=8$
16. $\boxed{4}+6=10$

Lesson Test 19

1. $4-1=3$
2. $6-5=1$
3. $8-8=0$
4. $1-0=1$
5. $2-1=1$
6. $4-3=1$
7. $7-0=7$
8. $9-1=8$
9. $5-4=1$
10. $9-8=1$
11. $6-6=0$
12. $10-1=9$
13. $7-6=1$
14. $6-1=5$
15. $3-3=0$
16. $0-0=0$
17. $\boxed{0}+7=7$
18. $\boxed{2}+1=3$
19. Ten minus one equals nine.
20. $4-0=4$ dimes

Lesson Test 20

1. $9-2=7$
2. $8-6=2$
3. $3-2=1$
4. $9-7=2$
5. $5-2=3$
6. $6-4=2$
7. $7-2=5$
8. $7-5=2$
9. $6-2=4$
10. $10-8=2$
11. $2-2=0$
12. $7-1=6$
13. $7+9=16$
14. $9+2=11$
15. $4+9=13$
16. $9+9=18$
17. $\boxed{9}+5=14$
18. $\boxed{9}+8=17$
19. $8-2=6$ children
20. $7-5=2$ points

Lesson Test 21

1. $11-9=2$
2. $10-9=1$
3. $14-9=5$
4. $17-9=8$
5. $12-9=3$
6. $13-9=4$
7. $18-9=9$
8. $16-9=7$
9. $15-9=6$
10. $9-9=0$
11. $9-2=7$
12. $8-6=2$
13. $5+8=13$
14. $8+7=15$
15. $4+8=12$
16. $800+100=900$
17. $\boxed{8}+6=14$
18. $\boxed{3}+8=11$
19. $11-2=9$ cars
20. $5-3=2$ dollars

Lesson Test 22

1. 12 − 8 = 4
2. 15 − 8 = 7
3. 13 − 8 = 5
4. 14 − 8 = 6
5. 16 − 8 = 8
6. 10 − 8 = 2
7. 17 − 8 = 9
8. 9 − 8 = 1
9. 15 − 8 = 7
10. 12 − 9 = 3
11. 16 − 9 = 7
12. 14 − 9 = 5
13. 4 + 4 = 8
14. 7 + 7 = 14
15. 6 + 6 = 12
16. 300 + 300 = 600
17. $\boxed{7}$ + 2 = 9
18. $\boxed{6}$ + 1 = 7
19. 11 − 8 = 3 chapters
20. 17 − 8 = 9 dollars

Lesson Test 23

1. 10 − 5 = 5
2. 2 − 1 = 1
3. 14 − 7 = 7
4. 60 − 30 = 30
5. 4 − 2 = 2
6. 16 − 8 = 8
7. 12 − 6 = 6
8. 8 − 4 = 4
9. 18 − 9 = 9
10. 17 − 8 = 9
11. 12 − 8 = 4
12. 15 − 9 = 6
13. 4 + 5 = 9
14. 9 + 1 = 10
15. 3 + 7 = 10
16. 8 + 2 = 10
17. $\boxed{5}$ + 5 = 10
18. $\boxed{7}$ + 9 = 16
19. 6 − 3 = 3 hours
20. 12 − 6 = 6 miles

Lesson Test 24

1. 10 − 3 = 7
2. 10 − 5 = 5
3. 10 − 2 = 8
4. 10 − 4 = 6
5. 10 − 6 = 4
6. 10 − 8 = 2
7. 10 − 7 = 3
8. 10 − 1 = 9
9. 16 − 8 = 8
10. 6 − 3 = 3
11. 14 − 7 = 7
12. 18 − 9 = 9
13. 5 + 4 = 9
14. 7 + 3 = 10
15. 2 + 7 = 9
16. 3 + 6 = 9
17. $\boxed{8}$ + 1 = 9
18. $\boxed{4}$ + 5 = 9
19. 10 − 4 = 6 years
20. 10 − 7 = 3 cards

Unit Test III

1. 13 − 9 = 4
2. 11 − 8 = 3
3. 11 − 9 = 2
4. 14 − 9 = 5
5. 3 − 1 = 2
6. 8 − 4 = 4
7. 7 − 1 = 6
8. 4 − 0 = 4
9. 6 − 2 = 4
10. 9 − 9 = 0
11. 9 − 8 = 1
12. 10 − 5 = 5
13. 3 − 2 = 1
14. 7 − 2 = 5
15. 15 − 9 = 6
16. 2 − 1 = 1
17. 16 − 9 = 7

18. 6 − 3 = 3
19. 10 − 9 = 1
20. 8 − 2 = 6
21. 18 − 9 = 9
22. 5 − 1 = 4
23. 17 − 8 = 9
24. 13 − 8 = 5
25. 7 − 5 = 2
26. 8 − 7 = 1
27. 17 − 9 = 8
28. 14 − 8 = 6
29. 5 − 3 = 2
30. 0 − 0 = 0
31. 9 − 2 = 7
32. 7 − 6 = 1
33. 14 − 7 = 7
34. 12 − 9 = 3
35. 6 − 4 = 2
36. 12 − 8 = 4
37. 6 − 5 = 1
38. 16 − 8 = 8
39. 5 − 4 = 1
40. 2 − 2 = 0
41. 6 − 0 = 6
42. 8 − 8 = 0
43. 8 − 6 = 2
44. 12 − 6 = 6
45. 4 − 3 = 1
46. 9 − 7 = 2
47. 13 − 8 = 5
48. 7 − 0 = 7

Lesson Test 25

1. 9 − 4 = 5
2. 9 − 7 = 2
3. 9 − 6 = 3
4. 9 − 0 = 9
5. 9 − 2 = 7
6. 9 − 3 = 6
7. 9 − 1 = 8
8. 9 − 8 = 1
9. 10 − 6 = 4
10. 16 − 8 = 8
11. 10 − 7 = 3
12. 14 − 7 = 7
13. 4 + 3 = 7
14. 3 + 5 = 8
15. 8 + 5 = 13
16. 9 + 4 = 13
17. $\boxed{4}$ + 1 = 5
18. $\boxed{4}$ + 3 = 7
19. 9 − 4 = 5 people
20. 9 − 6 = 3 people

Lesson Test 26

1. 7 − 3 = 4
2. 8 − 5 = 3
3. 7 − 4 = 3
4. 8 − 3 = 5
5. 9 − 5 = 4
6. 9 − 6 = 3
7. 15 − 8 = 7
8. 70 − 30 = 40
9. 12 − 9 = 3
10. 17 − 8 = 9
11. 14 − 7 = 7
12. 9 − 2 = 7
13. 7 + 4 = 11
14. 6 + 7 = 13
15. 7 + 5 = 12
16. 9 + 7 = 16
17. $\boxed{8}$ + 7 = 15
18. $\boxed{5}$ + 3 = 8
19. 7 − 4 = 3 inches
20. 8 − 3 = 5 years old

Lesson Test 27

1. 16 − 7 = 9
2. 13 − 7 = 6
3. 15 − 7 = 8
4. 11 − 7 = 4
5. 12 − 7 = 5
6. 7 − 3 = 4
7. 8 − 5 = 3

8. $70 - 40 = 30$
9. $8 - 3 = 5$
10. $9 - 5 = 4$
11. $10 - 6 = 4$
12. $16 - 8 = 8$
13. $5 + 6 = 11$
14. $6 + 9 = 15$
15. $8 + 6 = 14$
16. $6 + 7 = 13$
17. $\boxed{7} + 4 = 11$
18. $\boxed{9} + 9 = 18$
19. $15 - 7 = 8$ dollars
20. $13 - 7 = 6$ birds

Lesson Test 28

1. $13 - 6 = 7$
2. $15 - 6 = 9$
3. $11 - 6 = 5$
4. $14 - 6 = 8$
5. $15 - 7 = 8$
6. $8 - 3 = 5$
7. $13 - 7 = 6$
8. $70 - 30 = 40$
9. $9 - 7 = 2$
10. $8 - 1 = 7$
11. $15 - 9 = 6$
12. $11 - 8 = 3$
13. $7 + 5 = 12$
14. $5 + 9 = 14$
15. $8 + 5 = 13$
16. $6 + 5 = 11$
17. $\boxed{6} + 6 = 12$
18. $\boxed{9} + 4 = 13$
19. $11 - 6 = 5$ plates
20. $13 - 6 = 7$ toys

Lesson Test 29

1. $14 - 5 = 9$
2. $12 - 5 = 7$
3. $13 - 5 = 8$
4. $11 - 5 = 6$
5. $13 - 6 = 7$
6. $12 - 7 = 5$
7. $11 - 6 = 5$
8. $40 - 20 = 20$
9. $16 - 7 = 9$
10. $14 - 6 = 8$
11. $13 - 7 = 6$
12. $15 - 6 = 9$
13. $7 + 4 = 11$
14. $4 + 9 = 13$
15. $9 + 3 = 12$
16. $8 + 3 = 11$
17. $\boxed{5} + 6 = 11$
18. $\boxed{4} + 8 = 12$
19. $12 - 5 = 7$ words
20. $14 - 5 = 9$ pages

Lesson Test 30

1. $12 - 4 = 8$
2. $11 - 3 = 8$
3. $13 - 4 = 9$
4. $12 - 3 = 9$
5. $11 - 4 = 7$
6. $12 - 5 = 7$
7. $11 - 7 = 4$
8. $80 - 50 = 30$
9. $11 - 5 = 6$
10. $15 - 7 = 8$
11. $14 - 6 = 8$
12. $12 - 7 = 5$
13. $13 - 5 = 8$
14. $15 - 6 = 9$
15. $8 + 4 = 12$
16. $6 + 6 = 12$
17. $7 + 7 = 14$
18. $9 + 2 = 11$
19. $13 - 5 = 8$ answers
20. $11 - 3 = 8$ cookies

Unit Test IV

1. $9 - 4 = 5$
2. $7 - 4 = 3$
3. $11 - 3 = 8$

4. $11 - 6 = 5$
5. $13 - 7 = 6$
6. $12 - 7 = 5$
7. $13 - 5 = 8$
8. $11 - 4 = 7$
9. $9 - 6 = 3$
10. $7 - 3 = 4$
11. $11 - 5 = 6$
12. $13 - 6 = 7$
13. $12 - 4 = 8$
14. $8 - 3 = 5$
15. $8 - 5 = 3$
16. $11 - 7 = 4$
17. $12 - 3 = 9$
18. $15 - 7 = 8$
19. $14 - 5 = 9$
20. $9 - 3 = 6$
21. $15 - 6 = 9$
22. $16 - 7 = 9$
23. $9 - 5 = 4$
24. $14 - 6 = 8$
25. $12 - 5 = 7$
26. $13 - 4 = 9$

Final Test

1. $10 - 3 = 7$
2. $7 + 3 = 10$
3. $8 - 4 = 4$
4. $4 + 7 = 11$
5. $9 - 6 = 3$
6. $9 + 9 = 18$
7. $12 - 7 = 5$
8. $8 + 7 = 15$
9. $15 - 9 = 6$
10. $12 - 4 = 8$
11. $5 + 3 = 8$
12. $13 - 6 = 7$
13. $10 - 5 = 5$
14. $7 + 6 = 13$
15. $3 + 6 = 9$
16. $11 - 8 = 3$
17. $8 + 5 = 13$
18. $4 + 9 = 13$
19. $17 - 9 = 8$
20. $14 - 5 = 9$
21. $3 + 8 = 11$
22. $13 - 9 = 4$
23. $5 + 7 = 12$
24. $16 - 7 = 9$
25. $9 + 3 = 12$
26. $11 - 6 = 5$
27. $15 - 8 = 7$
28. $7 + 4 = 11$
29. $5 + 6 = 11$
30. $8 + 7 = 15$

Glossary

A-C

addend - a number that is added to another

Associative Property of Addition - a property that states that the way terms are grouped in an addition expression does not affect the result

circle - a simple closed curve with points that are all the same distance from the center

Commutative Property of Addition - a property that states that the order in which numbers are added does not affect the result

D-M

decimal system - a number system based on ten, also called *base ten*

decompose - to separate a number into parts

difference - the result of subtracting one number from another

equation - a mathematical statement that uses an equal sign to show that two expressions have the same value

minuend - a number from which another is to be subtracted

minus - decrease by subtraction

P-S

place value - the position of a digit which indicates its assigned value

plus - increase by addition

rectangle - a quadrilateral with two pairs of opposite parallel sides and four right angles

skip counting - counting forward or backward by multiples of a number other than one

square - a quadrilateral in which the four sides are perpendicular and congruent

subtrahend - a number to be subtracted from another

sum - the result of adding numbers

T-Z

triangle - a polygon with three straight sides

unit - the place in a place-value system representing numbers less than the base

unknown - a specific quantity that has not yet been determined, usually represented by a letter

Master Index for General Math

This index lists the levels at which main topics are presented in the instruction manuals for *Primer* through *Zeta*. For more detail, see the description of each level at mathusee.com. (Many of these topics are also reviewed in subsequent student books.)

Alpha Index